W9-BMZ-463

FRUITS & BERRIES OF THE PACIFIC NORTHWEST

*Green orchards contrast with basalt hill below Vantage on the
Columbia River.*

FRUITS&BERRIES

of the Pacific Northwest

What could be more delicious than fresh fruit?

DAVID C. FLAHERTY
and SUE ELLEN HARVEY

ALASKA NORTHWEST PUBLISHING COMPANY

Afternoon light silhouettes a fruit tree in British Columbia's Okanagan.

Copyright ©1988 by Alaska Northwest Publishing Company. All rights reserved. No part of this book may be reproduced or transmitted in any form or by any means, electronic or mechanical, including photocopying, recording or by any information storage and retrieval system, without written permission of Alaska Northwest Publishing Company.

Photographs by David C. Flaherty except as noted
Cover and book design by Shawn Lewis

Alaska Northwest Publishing Company
130 Second Avenue South
Edmonds, WA 98020

Printed in U.S.A.

Library of Congress Cataloging-in-Publication Data

Flaherty, David C.
    Fruits & Berries of the Pacific Northwest / David C. Flaherty and Sue Ellen Harvey.
        p.   cm.
    Bibliography: p.
    Includes index.
    ISBN 0-88240-328-1
    1. Fruit-culture—Northwest, Pacific. 2. Berries—Northwest, Pacific. 3. Fruit—Northwest, Pacific—Varieties. 4. Berries-Northwest, Pacific—Varieties. I. Harvey, Sue Ellen. II. Title. III. Title: Fruits and berries of the Pacific Northwest.
SB355.F57 1988

634′.09795—dc19   87-37425
                                                                        CIP

Contents

Binful of ripe apples offers plenty of tasty choices.

Roads through vineyards create pleasant geometry in the lower Yakima Valley.

With Thanks to Many . . .

When we started on this publication, we knew we would need considerable help. What we did not know, however, was that the people we would meet in the orchards, the vineyards, the packing houses, the processing plants, and the agricultural extension services would be so willing to be of assistance.

They were great!

It is difficult to single out any one individual as contributing the most to the project. However, we should start any such list with Robert Henning, Alaska Northwest's publisher, who initiated the concept of the book. We greatly appreciate the opportunity that he provided.

Then, chronologically speaking, there were Sherrill Carlson of Washington State University, who got us started; Grady Auvil of Orondo, Washington, who gave us more time than he had to spare; and James Ballard of Yakima, who furnished us with the benefits of his long experience with fruit growing and growers.

Soon thereafter there were Tom and Sue Perkins of Sedro Wooley; Kay Simon and Clay Mackey of Prosser; Gordon Saxton of Caldwell, Idaho; Bob Spanjer of Cashmere, Washington; and Greg Taylor of the Aplet/Cotlet organization, who opened our eyes to still more dimensions of the Pacific Northwest fruit industry.

Later on, numerous individuals sat still for an interview or took us on tours of their orchards and facilities. This group included Jake and Anne Van Westen of Naramata, British Columbia; Ken Severn of the Washington State Fruit Commission; Lydia Hall of Portland, Oregon; Charles and Julie Swanson of Corvallis, Montana; Jack Watson, grape specialist for Washington State University; Hal Tanaka of Dayton, Oregon; and Sharon Hall, sister of the famed "Fruit Stand Tiny" of Cashmere.

Earl Otis and Jo Ann Robbins of the Western Washington Research and Extension Center went more than a country mile to help us find specimens of small fruits to photograph. Ray Schneider, grape vine grower of Grandview, Washington, proved to be an outstanding resource, as did his neighbor, Paul Darby.

Toward the end of the project, Peggy Flaherty put in long hours at the keyboard, for which the authors are most grateful. Ballard also did double duty by reviewing the manuscript. He was joined in this task by Fred Westberg, retired manager of the Washington Fruit Commission, and Harold Rogers, former editor of *Western Fruit Grower*. Maureen Zimmerman, our editor, did her best to help us keep the photos, copy, and information organized.

Finally, there were the many growers who tramped down their orchard rows to show us their prize specimens. Their names may have eluded our memories but their courtesies will be long remembered.

The errors of omission and commission are ours, of course. We hope there are not too many.

Until we looked close-up through a camera lens at the colorful blushes on apples, pears, peaches, and all the other varieties in this book, we never realized how truly beautiful fruit can be. And to be honest, we did our share of taste sampling as we went along. How good it was!

David C. Flaherty and Sue Ellen Harvey

These dew-drenched Granny Smiths represent one of the most popular apple varieties grown in the Northwest.

It's So Good! Bright red to

yellow to green apples that yield up their goodness with sharp crunches; drippy-sweet peaches that blush in vibrant shades of red on gold; aromatic pears that dissolve in the mouth; plump berries in colorful shades of blue, red, purple, and black; clusters of pale to dusky grapes that saturate the taste buds — one of life's major delights has to be fruit. Whether it is a Granny Smith apple, a Golden Jubilee peach, or a rosy Rainier cherry, it looks good, it tastes good, and it's good for you.

There is little else in this world that has so many pluses and so few negatives as fruit. It is little wonder that the writers of the early parables cast fruit in the role of the supreme temptation. Would Adam have succumbed to the offer of a flapping fish or a squawking chicken?

Fortunately, fruit trees and vines grow most everywhere in the world except for areas of severe cold. And even more luckily for the lovers of fruit, most varie-

ties produce staggering amounts per tree or vine. Fruit purchased off-season may seem dear, but an apple or peach purchased when picking is in full swing is always a good bargain in both pleasure and nutrition.

Not only is fruit a delight to consume — it can be enjoyable to grow. Most fruit begins in a cloud of fragrant blossoms on the tree, a visual bonus.

This book has been prepared to give the reader general information on the kinds of fruit suitable for growing in most parts of the Pacific Northwest.

Included is an overview of planting and caring for fruit trees and vines. Persons interested in growing fruit should supplement the information in this book by visits to local nurseries and agricultural agents.

Also in the book is information on the fascinating fruit industry and stories about some of the people who have devoted their lives to keeping supermarket bins filled with attractive and healthy fruit.

So read on, and bon appetit!

The Varieties
of Fruits and Berries

Irrigated vineyards stretch over flat plain near Prosser.

The Pacific Northwest has it all — river valleys that twist and turn enroute to the ocean, broad prairies bearing thick coats of grass and shrub, mountains and ridges clad with green conifers, deserts spotted with sagebrush and feathery-headed bunchgrass, damp coastal reaches washed with frequent rains — the list goes on and on. This almost bewildering diversity makes the region an enchanting place to live, but it also makes it difficult to devise an uncomplicated set of recommendations for choosing varieties of fruit.

About the only general statement that can be made is that conditions are generally different on the west side of the Cascade Mountains than on the east side. However, there are many sub-regions and elevations that are different from their geographical neighbors. Thus what might grow well in the remote mountain-girt Illinois Valley of southwestern Oregon could be

a sickly performer along the bottoms of the Yamhill, south of McMinnville, Oregon. Then in far-off Montana, Red McIntosh can thrive in the Bitterroot Valley south of Missoula yet might have a difficult time in the same state's Clark Fork Valley farther to the north.

Pinpointing all of these individual climatic regions is beyond the scope of this publication. Most varieties are marked with an "East"/"West" to show their general suitability for orchards east or west of the Cascade Mountains. However, it is strongly suggested that this information be supplemented with recommendations from local nurseries, county agricultural extension agents, and persons in your area already growing fruit. You may find that although you live east of the Cascades, you can do well with a "West" variety in your orchard.

Apples

AKANE (East/West)

Developed at the Amori Experiment Station in Japan, Akane was introduced into the United States in 1970. Derived from the Jonathan, the Akane has cherry red skin with crisp, white, juicy flesh.

Although it is suited for warmer climates than the Pacific Northwest, the Akane lasts well on the tree, which bears annually at an early age. It does not store well but is an excellent dessert variety, with a slightly acid flavor.

CHEHALIS (West)

This apple, a Golden Delicious type, is large in size with a greenish-yellow to full yellow skin that sometimes has a pink blush. Its flesh is white to cream in color, crisp in texture, juicy and slightly acid. Chehalis is not particularly aromatic, but it is an excellent dessert apple that keeps well in cold storage. It is a good choice for orchards west of the Cascade Mountains.

Chehalis keeps well in cold storage and is resistant to scab, but it bruises easily.

CORTLAND (East/West)

The Cortland is a large, roundish apple with a dark red skin that is underlaid with stripes. Its white flesh is slow to turn brown, making it attractive in salads. The flesh is crisp, tender, juicy and sub-acid.

A cross between the McIntosh and Ben Davis, the Cortland is borne early in the year in a heavy crop, on a hardy tree. The attractive fruit, which originated in New York, handles well.

Many apple growers rate the Cortland even better than the McIntosh for both eating and cooking. The fruit stores well, but the tree is susceptible to powdery mildew. Temperatures above 100°F. in the weeks just prior to harvesting will ruin the fruit.

EARLY CORTLAND (East/West)

The appearance of this fruit is similar to Cortland but the flavor is more tart. Splashed with red striping over a light green ground color, Early Cortland is large, uniform and round in shape. It ripens approximately thirty days earlier than Cortland, and hangs well on the tree.

CRITERION (East)

This good-tasting yellow apple with a faint red blush has the traditional shape of a Red or Golden Delicious but it beats them for size,

Akane

Criterion

Golden Delicious

Red Delicious

Gala

Granny Smith

hands-down. Some Criterions heft out at two pounds or better, making them a good choice for gift packages where size is important.

Criterions won five out of six taste tests conducted by *Sunset* magazine at the Puyallup State Fair in 1983. Their flavor is described as mildly sweet and aromatic.

The Criterion emerged as a chance seedling from a Wapato, Washington orchard. Its parentage is unknown, but speculation is that the Criterion may be part Winter Banana, Red Delicious, or perhaps Golden Delicious.

DELICIOUS, GOLDEN (East/West)

Along with Red Delicious, the Golden is one of the leading commercial varieties in the Pacific Northwest. A large, yellow-skinned fruit with an occasional pink blush, Golden Delicious is an all-purpose apple: excellent fresh, or in desserts, salads, and sauces.

The apple itself is crisp, firm, juicy, sweet, and aromatic. The trees bear young with a heavy load of high-quality fruit. There are numerous sub-varieties.

DELICIOUS, RED (East)

According to many experts, this is the most economically important apple grown in the United States, with many different cultivars and sports. It is said to be the most widely grown apple in the world. It has a shiny, solid to striped, red waxy skin with creamy, very juicy flesh. Red Delicious trees bear early and heavily. The fruit keeps well in controlled atmosphere (CA) storage.

The Red Delicious apple is medium-sized, long and tapering. It is highly aromatic, firm, and sweet.

However, these qualities, and its appearance, will vary from strain to strain.

EMPIRE (East/West)

A cross between a McIntosh and a Delicious, Empire is medium in size and round in shape. It has a dark red skin with a striped blush, and cream-colored flesh. The Empire is crisp and juicy, has an excellent flavor, and is at its best in the fresh markets.

Empire originated in New York in 1966, is aromatic and sub-acid, but is regarded as unsuitable for processing use.

GALA (East/West)

Originating in New Zealand, Gala has several different sub-varieties. All are characterized by bright scarlet stripes over a yellow background. Aromatic, with a semi-sweet flavor, Gala is a dessert-quality apple with firm, fine-textured, yellow-white flesh that keeps well. Gala is a cross between Golden Delicious and Cox's Orange. Red strains of Gala include Imperial, Regal, and Royal.

EARLY GOLD (East/West)

Discovered in Selah, Washington, this variety resembles the Golden Delicious. Early Gold is considered by some to be one of the best early yellow apples and a good replacement for the Lodi variety.

GRANNY SMITH (East)

This increasingly popular green-colored apple originated in Australia. Some orchardists are forecasting that the Granny Smith will replace the Golden Delicious in popularity in the Pacific Northwest.

A late-maturing apple, the Granny is large in size, firm in flesh, bruise-resistant, and slightly tart. It is not recommended for planting west of the Cascades except in Oregon's Willamette Valley.

GRAVENSTEIN
(East/West)

The Gravenstein, an apple that has been around for many years, is slow to begin bearing, but has an outstanding flavor and is good for both cooking and eating fresh. It is medium to large in size, roundish to irregular in shape, with red stripes over a light green background.

The flesh is fine-textured, crisp, firm, and juicy, but the shelf life is very short. The trees are vigorous, upright, and spreading.

A variant, the Red Gravenstein, which is more highly colored, does well west of the Cascade Mountains. An improved red strain, it is especially good for pies and applesauce.

HAWAII (East/West)

A Golden Delicious-type apple with sweet and juicy flesh, Hawaii is a good keeper. However, it is susceptible to scab and slightly susceptible to bitter-pit. A Gravenstein-Yellow Delicious cross, Hawaii is a good substitute for Golden Delicious in west-side orchards.

The tree is tall, vigorous, and productive.

IDARED (West)

A popular variety, Idared is a handsome, large, bright red fruit, ripening just ahead of Red Rome. It exhibits white, firm flesh that is mildly acid. A late-keeping dessert and processing apple that is aromatic, Idared is tart at harvest time with a flavor that continues to improve with storage.

Idared, which resembles the Jonathan, is called by some observers a "strikingly beautiful" apple. Named by the Idaho Agricultural Experiment Station in Moscow, Idaho in 1935, it has medium-thick, waxy skin. The trees bear young, but are susceptible to fire blight and powdery mildew.

JONAGOLD (East/West)

A cross between the Golden Delicious and Jonathan, Jonagold is regarded as one of the best all-purpose apples. Excellent for eating out-of-hand, as well as for cooking, the Jonagold stores well and is large, crisp and juicy.

The trees are sturdy, productive, and good for growing in all districts of the Pacific Northwest. They are fairly susceptible to cold damage but no more so than the Delicious.

Jonagold has bright red stripes over a yellow ground. Jona-Go-Red is an all-red strain.

JONAMAC (East)

As can be deduced from its name, this apple is a cross between the Jonathan and McIntosh varieties. Growers, however, consider it a distinct improvement over both parents. Considered a dessert apple, the Jonamac has firm, crisp flesh that is excellent in flavor when allowed to tree-ripen.

The Jonamac stores well but is susceptible to scab and mildew. The apple has a greenish ground color overlaid with a streaky red blush.

JONATHAN (East/West)

Jonathan apples, which originated in New York before 1826, are moder-

Hawaii

Jonagold

Jonamac

Jonathan

Lodi

McIntosh

Melrose

ately tart with crisp, very flavorful flesh. Small to medium in size, the Jonathans have red stripes over a yellow background. The tree itself is small with low vigor, and is susceptible to mildew. Some of the most popular red strains of the Jonathan include Blackjon, Jonee, Jon-A-Red, and Super-Jon.

KING (West)

A "pioneer apple" in western Washington state, the King has red stripes over a greenish-yellow ground color. Good for cooking, eating, and storing, King has very large fruit that is richly flavored and sweet. The somewhat weak and slow-growing tree lacks winter hardiness.

LIBERTY (West)

An excellent commercial apple that originated in New York, the Liberty is resistant to most major apple diseases. It is medium to large in size, with a red blush, and has crisp, juicy, sweet flesh that is slightly coarse. The Liberty is similar to a McIntosh in hardiness.

LODI (East/West)

An early yellow apple, the Lodi has only a moderate shelf life but possesses a rich, spritely flavor. Good for sauce and pies, the Lodi's crisp flesh is tart and slightly acid.

The tree is large and productive but bears every other year. The Lodi originated in Geneva, New York, and is sometimes called a "Big Transparent."

MACOUN (East/West)

Originated in Geneva, New York, the Macoun is similar to the McIntosh, with a very dark red color at

maturity. A good dessert-type apple with high-quality flesh, the Macoun is best for roadside and local markets.

The trees bear heavily every other year unless thinned and are upright in growth with long, lanky branches. The ripe fruit drops off the tree.

McINTOSH (East/West)

The McIntosh, which originated in Ontario, Canada around 1800, grows on a winter-hardy, large, vigorous tree that needs special care and thinning when grown west of the Cascades. The best-quality McIntoshes are grown in cold climates. Summer temperatures above 100°F. are bad news for this variety.

The apples are medium in size, nearly round, and have a yellow ground color overlaid with a bright red blush.

The moderately soft and juicy flesh is good for cooking and for eating out-of-hand.

There are many sub-varieties of McIntosh.

MELROSE (West)

The Melrose, which originated in Ohio, is generally medium in size with yellow skin that is dotted with bright red blotches. (Western Washington orchardist Tom Perkins — page 56 — calls the Melrose his favorite ugly apple!) The flesh is firm, juicy and white, and slightly acid in flavor. Excellent dessert and cooking apples that store well, Melrose apples are good for roadside and local market sales.

A cross between the Jonathan and Delicious apples, Melrose can be a good choice for growers located west of the Cascades. The tree is productive but the variety is highly susceptible to scab and mildew infestations.

MUTSU (West)

A Japanese variety, the Mutsu is a cross between the Golden Delicious and the Indo. It matures after Golden Delicious and stores extremely well. The fruit itself is large and yellow-green in color with an orange blush. The flesh is mildly sub-acid, dense, and juicy.

Some orchardists consider the Mutsu to be a good replacement for the Golden Delicious because of the former's strong resistance to injury from chemical sprays. Others still consider the Golden Delicious superior.

The Mutsu does not shrivel in storage, bears heavily every other year, but may have bitter-pit.

NEWTOWN (East/West)

Called by some "the classic American apple," this round fruit is large in size with solid green skin and cream-colored, crisp, tart flesh. Ideal for eating fresh, baking and cooking, and for cider, Newtown keeps well. A New York variety that emerged in the early 1700s, Newtown is not planted much anymore.

NORTHERN SPY
(East/West)

Now sold mostly for processing, this old-time variety also is a good eating apple. Northern Spy trees bear large, oblong-shaped fruit flattened at the base. The color is bright red over a pale yellow skin. The yellow flesh is sub-acid, aromatic, and has a delicious, tart flavor.

PAULARED (East/West)

Originally from Michigan, Paulared colors early with a solid red blush and hangs well on the tree. Equally good for eating fresh, in sauces and in pies, the apples are medium in size, with very firm, non-browning flesh. Paulareds have a fair storage life.

ROME BEAUTY (East)

Rome Beauties are large in size and round in shape with red skins and greenish-white flesh. With a subtle flavor good for baking, Rome Beauties are not suited for high altitudes as they mature late.

There are numerous sub-varieties of this apple, which originated in Ohio in the nineteenth century. The trees are small and compact but produce heavily.

SPARTAN (East/West)

The Spartan, which originated in Summerland, British Columbia, is a cross between the McIntosh and the Newtown. Smaller than the McIntosh, the Spartan has solid, red-colored skin that is very dark, and crisp, white, juicy flesh. An excellent dessert apple, the Spartan keeps well in cold storage. The tree bears annually if thinned, grows vigorously, and produces heavily.

SUMMERRED (East/West)

This apple, a cross between the McIntosh and the Golden Delicious, originated in Summerland, British Columbia. The trees bear early and regularly; the apples hang well and do well in storage. The fruit is medium to large with bright red skin and tender flesh. The apples are tart until fully ripe.

TYDEMAN'S EARLY
(East/West)

Derived from the McIntosh, Tydeman's Early, with moderately

continued on page 10

Mutsu

Paulared

Rome Beauty

Spartan

PROFILE:
A Pioneer Apple Shipper

Sam Birch

"Every Monday early, start out for one of the orchard areas. Every Saturday at noon, head for home and the office back in Portland, Oregon.

"At least twice a year he would see every grower on his list. He would go to Hood River, Parkdale, and Dee in Oregon; White Salmon, Stevenson, Yakima, and Wenatchee in Washington; Weiser and Meiser in Idaho; and many other places.

"We drove many thousands of miles. He was happiest when on the road, meeting people."

Lydia Hall, now of Beaverton, Oregon, is describing the travels of her father, Sam Birch, a pioneer apple shipper of the Pacific Northwest.

"Sam did not drive. My sister Doris or I, who were in our teens and early twenties then, did the driving in a 1924 Nash touring car. The high-pressure tires in that big car took seventy-two pounds of air. A lot of muscle was needed to turn the steering wheel!"

Lydia says she will never forget those early-day trips with her father. "All but the main roads were just gravel or plain dirt. Some had some oil added to the dirt.

"The potholes would be filled with dust. You would not know a pothole was there until you hit it!"

The Beaverton resident especially remembers the driving in Montana. "Many miles were what we called corduroy roads. They were made out of small logs placed across the road. They were added when streams washed out the original roadbed."

The fruit industry was considerably different back in the twenties and thirties when Birch, a native of England, was taking the early orchardists' Jonathans, Winesaps, Rome Beauties, and Spitzenbergs on consignment. Birch dealt directly with the growers, as the apple grower associations were just getting started. "Sam generally returned better prices to the orchardists than the cash buyers," Lydia states. "The growers looked forward to his twice-yearly visits." By the 1920s, Birch's exports to the United Kingdom and other European markets became quite large.

Despite the growth of his export business, Birch felt that a handshake was enough to bind both parties, Lydia recalls. "Usually about lunch time, we would drive into the grower's place. Thus deals were often made at the lunch table. As we left, the grower and Sam would agree on the number of boxes, the varieties, and the sizes. They then would shake hands to bind the deal."

Birch kept the details of his arrangements with the growers mostly in his head. However, he did carry a small book with him in which to make some notes. Lydia recalls that her mother's first question for Sam when he returned to the office was, "Who did you sign up?" Alice, an efficient wife who ran the Birch export office located in downtown Portland while Sam was on the road, would then send out confirmations of the verbal agreements.

"In later years, she would see that Sam left on Monday with typed contracts," Lydia states. "But Sam still thought it wasn't necessary, seeing how he would shake hands with the growers."

Birch was the first to ship fruit through the

Panama Canal. "He shipped apples, pears, prunes, anything that was marketable overseas," states Lydia. Apples were shipped in refrigerated cars from points of origin to either New York for shipment overseas, or to Portland terminals for shipping through the Panama Canal.

Birch had come to the United States in 1918 for the T.J. Poupart Company, located in London's Covent Garden, to look into the eastern apple industry and to send a trial shipment to the United Kingdom. He shipped from the East Coast of the United States for three years, working mostly out of New York City.

In 1921 he came west to Oregon. "He fell in love with the countryside, the friendly people, and the beautiful apples that were appearing on the fruit markets," Lydia says. "One year later he brought his family to live in America, because by that time he was spending eleven months in America and one month in England."

The Birch exporting business prospered until World War II began to darken the world. In the early spring of 1933, Birch had noticed a change in his relationships with some firms located in Germany. Wanting to keep their business, he shipped a load of apples on the *Trondanger*, a Norwegian vessel, and accompanied them. Birch found the German company officials cold and casual, Lydia relates. Previously they had been extremely cordial. However, a good friend asked Sam to his house for lunch the following day. But he warned Sam to be very careful what he talked about. After lunch Sam and his friend went out for a quiet boat ride. The friend told him that there had been a change in the government with Adolf Hitler coming into power, that his office and house were bugged, and that his wife had had to dismiss her personal maid and take another who was sent to replace her, Lydia recalls.

Following this voyage, Birch's fruit exports continued to slack off. Gradually all of the Oregon pioneer's overseas shipping points were reduced by the growing conflict in Europe. In 1939, Sam Birch closed his export office. "He was a whiz kid," says Lydia. "But World War II was too big a problem even for Sam to overcome."

Label used by Sam Birch

Winesap

firm, creamy white flesh, is good for eating and cooking. The tree has long, lanky branches that can be attractive in an espalier treatment.

Also called Tydeman's Red, the fruit has a modest shelf life.

WINESAP (East/West)

Winesaps, all-purpose apple trees that originated in New York before 1817, are slow to bear. The fruit is mostly round in shape, medium in size, with striped red color over a yellow skin. They color early and keep well — but are losing popularity.

The yellow, very crisp but coarse flesh has a spritely flavor. Numerous red strains have been discovered and promoted.

WINTER BANANA (West)

The Winter Banana, which originated in Europe, produces a medium to large-sized fruit with pale yellow to pink skin. The flesh is tangy and aromatic. It's a good pollenizer but the flavor has a limited appeal.

YELLOW TRANSPARENT (East/West)

This variety, which originated in Russia a century ago, produces medium-sized fruit with clear yellow, thin skins and white, tender flesh.

The tart apples are good for cooking if harvested when greenish-yellow in color. However, they bruise easily and drop from the tree when ripe. They do not remain edible long, once ripe.

Apricots

Goldrich

BLENHEIM (ROYAL BLENHEIM) (East)

The Blenheim was once considered to be the best drying, eating, and canning apricot. However, it has been superseded by other varieties.

Popular in Europe, the Blenheim will not tolerate heat over 90°F. It is not a good choice for the British Columbia interior.

Yellow to yellow-orange in color, the Blenheim has firm, sweet flesh. Many homemakers can this variety with the pit intact.

CHINESE (East)

An early-bearing and heavy-producing variety with good flavor and texture, Chinese does well in colder climates but not in the British Columbia interior.

GOLDCOT (East)

The Goldcot, which originated in Michigan, has a bright gold skin with orange flesh. The fruit is firm and spritely in flavor; the cold-resistant tree blooms late and produces heavily.

The Goldcot is popular for fresh use and for processing, including utilization in baby foods.

GOLDRICH (East)

Introduced by Washington State University, Goldrich is a large, oval fruit with bright, shiny orange flesh that is firm. The tree is vigorous, productive, and hardy.

JANNES (West)

The Jannes variety, which originated in Washington state in 1943, does well in cool, moist climates, has a good flavor, and produces large, golden-yellow fruit.

MOONGOLD (East)

A Minnesota native, this good-quality apricot is orange in color with tough skin and orange-yellow flesh. It is ranked poorly for growing in British Columbia.

MOORPARK (East)

Some observers still rate this apricot, which originated in England before 1668, as setting the standard of excellence for apricots. The skin of Moorpark is orange with a deep red blush covered by dots of brown and red. The flesh is orange with a good flavor; it's good for canning, but is very soft when ripe.

Since not all the fruit on the tree ripens at once, Moorpark is an excellent variety for home gardens. The trees are hardy, except in the most extreme climates, but light-bearing. Wenatchee Moorpark is a well-known sub-variety.

PERFECTION (East)

A relatively new Washington state native, this apricot has a light orange-colored skin without a blush. It is a good choice for orchards located in mild-winter areas. The tree is vigorous and is preferred for commercial plantings except in British Columbia.

RILAND (East)

Another Washington state variety, the Riland has a rich, plumlike flavor and coarse-textured, firm flesh. Maturing early, the Riland keeps well. The fruit is large in size, with a rather flat shape. The skin has a deep red blush over half of the fruit.

RIVAL (East)

This apricot, which originated in Prosser, Washington, has a light orange skin with a pronounced blush, and comes from a vigorous, productive, and leggy tree. The fruit is good for canning, but the skin turns brown. The flesh is low in acid, with a mild flavor.

SCOUT (East)

Developed at a Manchurian fruit experiment station, this fruit became popular in Manitoba, Canada and was then imported into the Pacific Northwest. The fruit is roundish to flat in shape and bronze in color. It is good eaten fresh, canned, or used in jams. The tree, as might be expected, is hardy.

SKAHA (East)

The Skaha is a hardy variety of apricot from British Columbia. A good replacement for the Blenheim in colder areas, it is rated fair to good in texture and flavor.

Moorpark

Rival

Perfection

Tilton

STELLA (East)

Regarded as a good choice for home gardens in cold areas, Stella is a productive bearer with good flavor and acceptable texture.

SUNGOLD (East)

Originally a Minnesota variety, the Sungold has medium-sized, rounded fruit with tender golden skin that is blushed with orange. The fruit is good eaten fresh or preserved; the flavor is mild and sweet.

TILTON (East/West)

This apricot is regarded as one of the best varieties for growing west of the Cascades. The tree is vigorous, a heavy bearer, and tolerates heat. Originally from California, the Tilton is excellent for home canning and has a fair flavor when eaten fresh, but a poor flavor when dried. It is the most important commercial apricot in British Columbia.

WENATCHEE (East)

Widely grown in British Columbia and Washington for eating fresh, the Wenatchee bears heavy crops on a regular basis. The fruit softens on the tree, and may have green shoulders.

Blackberries

AURORA (West)

Well-suited to the Pacific Northwest and mild coastal climates in general, the Aurora is an early blackberry with excellent flavor.

BOYSEN (BOYSENBERRY, THORNLESS BOYSENBERRY) (West)

Another berry that does well in mild coastal climates is the Boysen. Large and aromatic on a vigorous plant that is fairly thorny, the Boysen is reddish-black at maturity. The flavor is suggestive of raspberries.

CASCADE (West)

Good fresh or preserved, Cascade blackberries have unsurpassed flavor and do well in the Pacific Northwest.

DARROW (East)

The Darrow, originally from New York, is very hardy and thus suited to those areas of eastern Washington where the cold is not too intense. The fruit ripens over a long season.

THORNLESS EVERGREEN (West)

This berry has black fruit with large seeds, and is fair in quality. It

Boysen

does well in mild coastal climates and is a top commercial berry variety in Oregon.

LOGAN (THORNLESS LOGAN) (East/West)

Dusty maroon to reddish-colored fruit characterize this excellent, tangy variety that is best in pies or preserves. The fruit is also used for making wine. Logan berries are at their best in the Willamette Valley but can be grown in the Columbia River area and in eastern Washington with the aid of winter protection.

MARION (East/West)

Berry plants of this variety require winter protection when grown in eastern Washington. The medium to large, high-quality fruit with excellent flavor is great for canning, freezing, pies, and jams. The large bushes have numerous spines. The fruit is bright black.

OLALLIE (West)

The Olallie blackberry is not very cold hardy, thus is better suited for the warmer areas of the Pacific Northwest. The black, glossy fruit is larger than the Boysen.

Marion

Olallie

Blueberries (East/West)

Blueberries do best in the cool coastal climates of Oregon, Washington, and British Columbia. However, they can be grown elsewhere except in the alkaline soils of Alberta.

BERKELEY

Good for all coastal zones, the Berkeley has a light blue skin. Originally grown in New Jersey, the Berkeley's large fruit is excellent for pies or eating fresh.

BLUECROP

Another New Jersey blueberry, the Bluecrop is rather tart but stores well and is good for cooking. This variety is drought-tolerant.

BLUERAY

Blueray has large, firm, sweet fruit. The hardy plants, which originated in New Jersey, produce heavy yields.

COLLINS

A fairly hardy variety, Collins is a good choice for growers in the more inland regions.

Blueray

Jersey

Bing

Black Republican

Corum

EARLIBLUE

One of the best blueberries for all areas, Earliblue produces fruit that is large, flavorful, and good for desserts. The plant is hardy and moderately vigorous.

JERSEY

This widely grown variety bears medium-sized, light blue fruit in long, loose groupings.

STANLEY

The Stanley is regarded as being tasty, aromatic, and spicy. It is a widely recommended variety.

WEYMOUTH

Weymouth blueberries are round, large in size, and have dark blue skins. They are good for cooking.

Cherries, Sweet

ANGELA (East)

This cherry originated at Utah State University. Hardier than the Lambert, it bears medium to large fruit that has a good flavor.

BING (East/West)

The Bing has been, for many years, the standard by which all black sweet cherries are judged. It has a dark red to mahogany-colored skin with firm, very juicy flesh. Excellent for canning, Bings are also good to eat fresh.

Originating in Oregon around 1875, this cherry is a good selection for growers west of the Cascades. However, Bings can exhibit some problems in humid climates and rain will make them crack.

A good shipper, the Bing is the leading commercial sweet cherry in the western states. Low temperatures in the winter or spring can damage the tree and blossoms.

BLACK REPUBLICAN (East)

Sometimes called the Black Oregon, this cherry has firm flesh that is very dark red in color. Small and tart, it keeps well. It originated in Oregon around 1850.

BLACK TARTARIAN (East)

This large, bright, dark-colored cherry, which originated in Russia, is smaller than the Bing. It is an immense and early bearer, but softens quickly when picked.

CORUM (East/West)

A yellow cherry with a red blush and thick, sweet, firm flesh, Corum originated in Oregon. A good canning cherry, Corum is planted commercially in Oregon.

EARLY BURLAT (East/West)

Similar to the Bing variety, this large cherry is an early bearer that is medium to firm in texture.

GOLD (East)

A bright yellow cherry that originated in Nebraska, the Gold can withstand temperatures down to 30°F. Gold is used in the production of maraschino cherries. The tree and blossoms are regarded as being especially hardy. It is suggested for home gardens in cold areas.

LAMBERT (East/West)

The Lambert is a large, dark red, heart-shaped cherry of excellent quality. Sometimes referred to as the "connoisseur's cherry," it originated in Oregon around 1875. It grows on a hardy and vigorous tree with strong, upright growth that is resistant to late frost. A good shipper with a rich flavor, the Lambert is one of the best-known commercial varieties.

RAINIER (East/West)

Originating in Prosser, Washington in 1960, this large, high-quality cherry has a yellow skin with a red blush, and clear flesh and juice. Wonderful eaten fresh, the cherry is also very good for canning. The tree is very productive (it tends to over-bear), and is hardy.

ROYAL ANN (NAPOLEON) (East/West)

The Royal Ann is a very old French variety that still is used heavily in commercial candy making and as a maraschino cherry. Yellow with red cheeks, the Royal Ann also is excellent for eating fresh and for canning. The fruit can be very large in size and tends to double in hot climates. The tree and blossoms are somewhat tender.

SAM (East/West)

An early-season variety from Summerland, British Columbia that was developed in 1953, Sam is a black-skinned cherry with good resistance to cracking. It is satisfactory for canning. The vigorous tree bears heavily when mature.

STELLA (COMPACT STELLA) (East/West)

This large, dark red, heart-shaped cherry was developed in Summerland, British Columbia in 1968. The variety does very well in orchards west of the Cascade Mountains. The tree is productive but tender; the fruit is resistant to cracking and good for canning.

VAN (East/West)

Like the Bing, the Van has dark, shiny red fruit of excellent quality. The vigorous tree is a heavy bearer and relatively hardy, but is subject to some diseases west of the Cascade Mountains.

The Van was developed at Summerland, British Columbia in 1944 and is widely planted in British Columbia.

Lambert

Rainier

Royal Ann

Van

Cherries, Pie

Montmorency

There are two major categories of pie cherry: the Amarella, which has clear juice and yellow flesh, and the Morello, which has red juice and flesh.

ENGLISH MORELLO (East)

A good variety for northern growers as it is a late ripener, English Morello has slightly tart, dark red flesh. It's an excellent choice for pies. The tree is only moderately vigorous.

METEOR (East)

This Amarella cherry originated in Minnesota and is especially hardy, but also does well in mild climates. It has large, light red fruit and yellow flesh. The flavor is considered to be fair to good.

MONTMORENCY (East)

A popular tart cherry, the Montmorency (an Amarella) is clear red on the outside but with yellow flesh inside. Originally from France, it is a favorite for processing and for home use. The productive tree is hardy.

NORTH STAR (East)

Since the North Star originated in Minnesota in 1950, one might expect that the tree would have good winter hardiness — and it does. North Star cherries will hang on the tree for two weeks after ripening.

Good for pies, cooking, and freezing, the North Star, a Morello, is dark red to mahogany in color with juicy, meaty flesh. It's self-fertile and an excellent choice for a garden.

RICHMOND (EARLY) (East)

Although it is astringent when eaten fresh, the Richmond cherry (an Amarella) is good for jams, jellies, preserves, and pies. It has a bright red skin and a tart flavor. The tree is especially hardy.

Cranberries (West)

If you are interested in having more than the usual Thanksgiving and Christmas acquaintances with the bright red fruit of cranberries, it is helpful to have a bog or marsh on your property. Although the cranberry plant can be grown as an ornamental ground cover, you need a reasonably large area that is wet a good portion of the time if you want more than a bucketful.

Bog is flooded so cranberries can be gathered after they're knocked loose from vines. Photo courtesy of Dennis Brown.

Cranberries are native to the bogs and marshes of northeastern America, across the northern tier of states, and the southern provinces of Canada. Massachusetts is reported to be the leader among the states, with British Columbia producing the most in Canada. Pacific Northwest growers are reported to grow the more highly colored varieties, a major advantage in marketing.

Acid soil is a must, as is an ample supply of water for irrigating during dry periods and for flooding during cold periods. A good supply of cheap water also is necessary in those areas where self-propelled machines are used to knock the berries loose from the vines and into the water just covering the vines. The floating berries are then picked up on a conveyor belt and deposited in a hopper.

It takes several years for cranberry bogs to become established, with full production starting in the fourth year.

In Washington, most commercial cranberry operations are located near Ilwaco and Long Beach where the state's southwestern corner juts into the Pacific Ocean. Oregon's bogs are situated primarily in Clatsop, Coos, Curry, and Tillamook counties, all located where the ocean breezes can moderate the climate. The Fraser River Valley of British Columbia is home to that province's red-berried bogs.

The varieties most commonly found in the Pacific Northwest are McFarlin, Crowley, Bergman, and Stevens.

Cranberries

Currants (East/West)

Red currants

The four most popular varieties of red currants that do well in the Pacific Northwest are Perfection, Red Lake, Stephens #9, and Wilder. These are medium to large red berries that thrive in Oregon and the valleys of eastern Washington. Red Lake, Stephens, and a variety called Prince Albert are considered suitable for Alberta growers.

White Grace and White Imperial are the two most popular white currants grown in the Pacific Northwest. Cherry and Prolific Fay's are often grown in western Oregon. The black currants Willoughby and Boskoop are suitable for Alberta.

Shiny red or pale white currants are excellent raw material for jams and jellies.

Gooseberries (East/West)

Gooseberries

The most common gooseberry in western gardens is the Oregon Champion. As the name suggests, the berry originated in Oregon. It bears medium-sized fruit that is light green to yellow in color. The Fredonia, from the east coast of the United States, has large, dark red fruit and is well-suited to the Pacific Northwest.

Both the Pixwel, from North Dakota, and the Poorman are red berries from hardy plants that do well in the northwest. The Clark, which originated in Ontario, Canada, has large red berries. Pembina Pride, which has green berries when ripe, is more vigorous than the Pixwell and is thought to be a more satisfactory choice for Alberta.

Grapes, Table and Juice

There are dessert grapes, raisin grapes, juice grapes, and wine grapes — and some varieties that can do double or triple duty in these categories! And like apples, the number of varieties and sub-varieties seems to be endless.

Essentially there are two major types of grapes, the American *(Vitis labrusca)* and the European *(Vitis vinifera)*. However, grape growers for hundreds of years have been busily crossing the two, hoping to achieve the best characteristics from the different varieties.

What follows is a selection of the most commonly recommended varieties. Additional suggestions for your particular locality can be obtained from your state or province's agricultural extension service. A good tip is to observe what varieties are doing well in existing vineyards, backyard orchards, and gardens.

One enterprising regional grower, Ray Schneider of Grandview, Washington, is trying to promote the wider production of different varieties in the Pacific Northwest. At the time this book was prepared, Schneider had some three hundred different grapes on his Yakima Valley acreage. He reports excellent results — even without irrigation — with some table grapes, a sector of the fresh grape market long thought to be an exclusive province of California growers.

ALDEN (East/West)

This table grape bears large fruit with lots of body to delight the consumer. The red to purple, compact clusters are large in size. Vigorous pruning will help prevent over-cropping.

BLACK MONUKKA (East)

This hardy European variety produces reddish-black grapes that have only a few seeds. The medium-sized grapes, which are oval in shape, come in large clusters on the vine.

BRONX SEEDLESS (East)

A large-berried variety with greenish to light red-purple fruit, Bronx Seedless is noted for its juicy quality. This grape originated at a New York state research station.

CAMPBELLS EARLY (East/West)

The dark purplish berries ripen about ten days earlier than the standard Concord. Large clusters of sweet, juicy grapes are borne on moderately vigorous vines. Camp-

Alden

Black Monukka

Bronx Seedless

Canadice Seedless

Concord

Flame Seedless

Interlaken Seedless

bells Early is recommended for climates that are too cool for Concord.

CANADICE SEEDLESS (East/West)

Considered one of the hardiest seedless table grapes, Canadice bears tight clusters of pale yellow fruit with red blushes that are a delight to eat.

CONCORD (East)

The big juice grape — both bottled and frozen — of the Pacific Northwest ripens from mid-season to late. A distinctive aroma permeates the vineyard when the blue-black, seeded grapes are ripe. Viticulturists describe the Concord — an American variety — as having a "foxy" flavor. The widely grown vines, which are cold-tolerant, are very productive.

CSABA (East/West)

A hardy European variety with pale golden grapes, Csaba has a light flavor with overtures of the more richly flavored muscats.

DELAWARE (East/West)

Small, light red berries in small clusters characterize this American variety. It is considered a good choice for out-of-hand eating and for juice, as well as for making into wine. Delaware grapes ripen in early mid-season.

FLAME SEEDLESS (East)

Gorgeous to the eye and pleasing to the taste, Flame Seedless comes in long, cone-shaped clusters. Light pink in color, the sweet berries hang well on the vines.

FREDONIA (East/West)

Rated as one of the top black grapes, with berries larger than Concord's, the Fredonia vines are cold-tolerant. An American grape with thick skins, the Fredonia originated in New York.

HIMROD (East/West)

A white seedless grape, Himrod is an American variety whose vines are not very hardy. Low yields are common from this early-bearer.

INTERLAKEN SEEDLESS (East/West)

A rich-yellow-colored grape that matures early, the Interlaken Seedless has medium-sized, round berries with an excellent flavor. The vines are rated as moderately hardy. Interlaken is a New York product that resembles the popular Thompson Seedless.

NIAGARA (East)

An American variety, the Niagara is one of the most widely planted white grapes that is good for both juice and wine-making. It is considered to be moderately hardy. The berries ripen in mid-season.

REMAILY SEEDLESS (East)

One of the largest of the seedless varieties, Remaily is golden to pale purple in color. Large-sized clusters will need to be thinned. This vine is regarded as moderately hardy.

SCHUYLER (East/West)

The tough skins of these blue grapes cover juicy and sweet interiors. Originated in New York state, the American variety does not produce copious quantities of fruit. The vines can handle a moderate amount of cold.

SENECA (East/West)

A white grape with golden skin and a sweet flavor, Seneca is aromatic. The hardy vine, which originated in New York, bears its fruit very early in the grape season.

SUFFOLK RED (East/West)

Large pink to red berries are borne in medium-sized clusters. The flavor is regarded as excellent, with the vines being moderately hardy.

Suffolk Red

Grapes, Wine

Some of the varieties listed previously are used to make wine. However, the major wine grapes for the Pacific Northwest include those following. In general, grapes that are high in acidity with a moderate sugar content are suitable for dry or table wines. On the other hand, sweet or dessert wines require grapes with a high sugar content and moderate acidity.

CABERNET SAUVIGNON (East/West)

Known throughout the world as a top-quality red wine, cabernet sauvignon is usually associated with France's famed Bordeaux wine area and with California's Napa Valley, Sonoma and Monterey regions. In 1985, however, over 1300 tons of the small, nearly black grapes were crushed in Washington state wineries. The vines do well in a variety of soils, but over-fertile soils will downgrade the quality of the grapes.

CHARDONNAY (East/West)

Known under a variety of other names, the pale green to light golden grapes of this variety are popular with vineyard owners throughout the world — Europe, the west coast of the United States, South Africa, and Australia. Reported to take well to careful irrigation in dry areas, the vines are regarded as hardy but susceptible to powdery mildew. The white wine is not strongly aromatic.

CHENIN BLANC (East)

Chenin blanc grapes are borne in large clusters that are long and conical in shape. The buds open early, making the variety subject to early frost damage. Now being planted in a variety of soils, Chenin blanc white wine is often blended with the juice of other white grapes to produce ordinary table wines. Originally from France's Loire

continued on page 26

Cabernet Sauvignon

Chardonnay

Vineyard near Prosser has Horse Heaven Hills as a backdrop.

PROFILE:

Washington Wine

Washington's wine industry might be young, but it couldn't be healthier, according to Washington State extension agent Jack Watson, located in Prosser, Washington. A number of conditions account for this fact, Watson indicates. Imports have become more expensive due to changes in the value of the dollar. Imported wine is frequently not as additive-free as wine produced in the United States, although this is not a major factor.

But the popularity of Washington wine has as much to do with the climate and the young wine producers in the state as it does with the current conditions surrounding the import market. "Washington produces only about 1/50 of the wine produced by California, but Washington doesn't produce 'jug' wine," Watson says. "Our wines are doing very well in competition with California wines," he adds. "Recently four of five wines submitted from Washington won awards in competition with about fifty different California wines."

Watson cites the cool growing area as one reason for the excellent flavor of many Washington wine grapes. According to the enthusiastic extension agent, the coolness contributes to a more fruity and aromatic quality, and a greater richness, in the wine. Additionally, eastern Washington's well-known dryness helps prevent some moisture-related diseases that affect grapes in other areas.

"I think eventually the Yakima Valley of Washington could be like the Napa Valley in California — everything it takes is here. There's lots of potential!" maintains Watson. But he is quick to point out that grape-growing in Washington is not, as far as he can tell, just a "rich man's fancy." Central Washington is, after all, an agricultural region already and wine grapes are simply being added to the list of crops already being produced by experienced growers.

None of these factors accounts for Washington's growing success in the wine grape industry, however, as much as the hard work and commitment demonstrated by a handful of dynamic young vintners in the Yakima-Prosser area. Husband and wife team Kay Simon and Clay Mackey exemplify the energy and team effort it takes to create and nurture a young winery. "Sure, the industry has a certain amount of sex appeal and attracts a number of people with lots of money," says Clay, "but we're interested in doing it because it's all we've ever done."

Clay worked in his parents' Napa Valley vineyard while going to college. Kay graduated from the University of California at Davis with a degree in Fermentation Science in '76. Clay worked in the Napa Valley for eight years altogether. Both worked for one of Washington's largest wineries, Chateau Ste. Michelle. Starting their own winery was almost a foregone conclusion.

But the birth of Chinook Wineries was brought about by anything but chance. Long hours of careful planning and years of study went into it.

When Mackey and Simon began thinking about exactly the kind of winery they'd like to operate, and the kind of wine they'd like to produce, they discovered a number of things. First of all, they realized that their goal was to focus on the special and unique in blended wine grapes.

Even their name and the design of their label

Chinook wines

is a conscientious attempt to avoid the names and designs most commonly in use among wineries

Clay Mackey and Kay Simon

today. "You always see pictures of vineyards on wine labels," says Kay, "and we decided that we wanted something different for our label, something that would really stand out. We chose the name *Chinook* because it makes a strong, local identity statement, and we deliberately went with a very non-traditional country kitchen-type label.

"Our product mix is uniquely ours, too. We specialize in strictly dry, oak-aged wines that will appeal to certain people. We aren't trying to produce something for everyone."

Chinook Wineries, operating since 1984, produces a small number of high-quality wines from European wine grape cultivars: sauvignon blanc, Semillon, chardonnay and merlot are the major varieties. Chinook ships about one hundred cases a month of its three top sellers, mostly to restaurants and wine shops in the Seattle-Tacoma area.

About Chinook's young owners, Watson says, "They're a real force behind Washington's growing wine trade. And what's particularly interesting about Washington wineries, I think, is the degree of sharing that goes on. These new businesses are all struggling to get started and yet you see none of the trade secret business you do in larger wine-producing areas. Kay and Clay speak frequently at meetings of Washington wine growers, sharing their expertise with their fellow competitors. These people even share equipment and facilities. It's great!"

Kay is equally enthusiastic about the spirit of cooperation among Washington wine-grape growers and small wineries. Frequently asked to consult with grape growers, wineries and grape crushers, Kay is at home in the field or the laboratory. In a still-growing industry, one full of newcomers, Kay's solid expertise is recognized.

According to Watson, Washington wineries are profiting from a heavy demand for white wines, and white wine grapes are, generally speaking, hardier and less risky for growers than red wine grapes. "People who don't ordinarily drink wine," Watson said, "will prefer a white wine. The progression seems to be from soda pop to a wine cooler to a white wine. Usually, the real wine lovers branch out to experiment with red wines." He adds that there are swings in popularity back and forth from whites, to reds, to rosés, and that not all years are equally good for wine grapes, even in Washington. But, regardless of the whims of nature and changing tastes, one things is certain: one local winery, snuggled demurely next to a plum orchard off Interstate I-82, will be quietly going about the business of creating some unique, appealing products for wine experts and neophytes from Washtucna to — who knows? — New York, New York!

Watson explains that a ton of grapes will produce approximately 140 gallons of wine. Washington grape growers expected to produce in excess of 40,000 tons of wine grapes in 1987. "We are very optimistic about wine growing in Washington. The sales of Washington wines have increased steadily during a time when California wines have remained static. Right now our bottlenecks are marketing and distribution. When we get on top of those problems, we're off and running!"

Washington growers also achieved a record wine grape harvest in 1986, despite harsh weather. Snoqualmie Winery's Joel Kline said, speaking of the state's wine growers in general, "Everybody came in high. I was 20 percent over myself. It was the earliest harvest I had seen since 1974 when the whole crop was estimated at 750 tons."

Unfortunately for the region's vineyard operations, these back-to-back record production years filled the winery tanks to overflowing. Although the vintners of the Pacific Northwest have achieved recognition, most are small to medium in size.

It is the same problem that vexes many industries. Until large crops of wine grapes are assured, few vintners can afford to add costly tanks.

Marketing consultant to Oregon state's winemakers, Fred Delkin, was also optimistic about the 1986 wine grape crop. "Everybody is saying it's as good as 1983, which was a landmark year," Delkin said. He is employed by the industry-sponsored Oregon Wine Advisory Board. Oregon has about fifty operating wineries, and Delkin estimates that six new ones a year have gotten started in the last three years in Oregon.

In a book entitled *The World of Canadian Wine,* author John Schreiner, western editor of the *Financial Post,* maintains that although wine has been made commercially in Canada since 1860, the best Canadian wines have come from vineyards and wineries less than a decade old. According to Schreiner, Canadian wines are reaching world standards. He cites three major reasons for this: the addition of new wine grape varieties in vineyards, improved technology in wine production, and a change in the taste of Canadian wine consumers away from novelty wines to table wines.

Schreiner lists six major wineries in Canada: Andres, Bright's, Calona, Casabello, Chateau-Gai, and Jordan/Ste-Michelle. He also lists small commercial wineries: Barnes, Beaupre, Culotta, London, Mission Hill, Okanagan, and The Wines of Quebec, and includes estate wineries.

In the conclusion to his introduction, he indicates that, "The wines of Canada have chosen a difficult decade for their entry onto the world stage. All the older wine-growing nations have surplus production, available at bargain prices to the affluent consumers of North America. Nor are Canadian wines alone in clamoring for the recognition they deserve, for new vineyards are being planted from Texas to Japan, and noble grapes are replacing coarse ones in places as widely separated as New Zealand and Bulgaria. Winemaking technology has improved everywhere; even trained winemakers are in surplus supply. It is a tough, competitive environment. Just as vines produce some of the best vintages from barely hospitable soils, however, the best wines of Canada will be made in this time of challenge."

Merlot

Riesling

Valley, Chenin blanc wine is described as having a fresh, fruity flavor.

GEWURZTRAMINER (East/West)

The golden to pink grapes of this variety produce a strongly scented wine, the most famous bottlings of which come from Alsace, France. California vintners have been trying their hand with the variety but conditions in the Pacific Northwest are considered more favorable. Relatively low yields and susceptibility to spring frosts can give growers difficulty. Deep, loamy, fertile soils are recommended.

MERLOT (East/West)

The plump berries of this world-famous red wine variety come in large but loose clusters. Used in blends with cabernet sauvignon, merlot does especially well in the warm days and cool nights of the Columbia Basin. Vines are widely planted through much of the world but are susceptible to early cold weather.

PINOT NOIR (East/West)

The black grapes of Pinot noir are famed for producing the outstanding red Burgundy wines of France as well as the paler versions of other countries. Considered a difficult variety to grow, Pinot noir is reported to do well in Oregon. Prone to rot, the vines respond favorably to well-drained deep soils.

RIESLING (East/West)

Originally a product of Germany's steep Rhine and Mosel valleys, Riesling vines — and numerous sub-varieties — are now grown throughout the world. In Washington state, the tonnage of Riesling grapes crushed far exceeds any other grape. Regarded as hardier than most varieties to cold, the Rieslings do mature late, however. Well-drained soils that are low in fertility are regarded as the best planting sites but success has been achieved with plantings in a wide variety of soils. The wine produced from the Riesling grape is noted for its strong floral aroma.

SAUVIGNON BLANC (East)

Recently popular with would-be fashionable wine drinkers, the whitish-green grapes of sauvignon blanc are borne in long, slender bunches. Most winemakers blend sauvignon blanc with Semillon to cut the "flinty" taste of the former. Highly aromatic, sauvignon blanc's smell has been described as "cat's pee on a gooseberry bush"!

Originally a product of France's Bordeaux region, sauvignon blanc seems to have found a home in California's Napa Valley. In Washington state, the variety was sixth in 1985 white wine tonnage crushed.

SEMILLON (East/West)

In the humid air of France's Bordeaux region, if the weather conditions are just right for the growth of *Botrytis cinerea* (noble rot), an outstanding white wine will be possible. Under less favorable conditions, a very sweet wine will result from the pale green to golden berries of Semillon grapes. They're often combined with sauvignon blanc to gain the advantage of the latter's tartness. The vines of Semillon are moderately productive.

Kiwi (West)

This homely but likable fruit is rapidly gaining in popularity across the United States. A native of eastern Asia, it was first planted as a commercial fruit in New Zealand in the 1950s. The kiwi, which grows on vigorous, deciduous vines, is a fuzzy greenish-brown fruit about the size and shape of a large egg. It should be grown on a trellis, on an arbor, or as an espalier.

The bright emerald-green flesh of the kiwi has small, black, edible seeds that form a starlike geometric pattern when the fruit is sliced through the center. An attractive addition to salads, the kiwi flavor is mildly minty, but also reminds some people of strawberry, melon, and banana flavors.

Kiwis grow well in the Puget Sound area and can be stored in the refrigerator up to four months. When planting kiwis, growers must plant at least one male and one female plant. The most popular varieties of female kiwi are: Hayward, Chico, Vincent, Monty, Bruno, and Abbott. The male kiwi plant is simply called the male kiwi.

Kiwi

Flavortop

Nectarines

FANTASIA (East)

A California product, Fantasia is bright red over yellow in color with freestone flesh. The tree is vigorous.

FLAVORTOP (East)

Also a freestone, Flavortop is colored bright red and yellow. The flesh has an excellent and distinctive flavor and ships well. The tree is tender but vigorous.

REDCHIEF (East)

A nearly round fruit, Redchief keeps well. It has a bright red skin with white freestone flesh.

RED GOLD (East)

The most commonly planted variety in Washington state, Red Gold has excellent storage and shipping qualities, fair hardiness, and very firm flesh; it is crack resistant.

Red Gold

Sunglo

Daroga Red

Dixiered

Elberta

A freestone, the Red Gold bears very large fruit with yellow flesh and a glossy red skin. The tree is vigorous.

SUNGLO (East/West)

This red-skinned freestone nectarine has golden flesh that keeps its flavor, whether this fruit is frozen or canned. Large in size, the Sunglo nectarine ripens after the Redhaven peach.

The tree is hardy and vigorous in production and growth habit.

Peaches

CANADIAN HARMONY (East)

A very productive freestone, Canadian Harmony must be picked before its full color has been reached if the fruit is to be shipped a long distance to market. It is a medium-firm fruit whose full ground color develops late in the ripening process. It is not recommended as a canning peach.

CHAMPION (East)

An old favorite, Champion was discovered in Illinois around 1880 as a seedling. Many consider it to be the best white peach for the middle of the season. The flesh is tender, melting, and juicy, with a sweet, delicate flavor. The tree is vigorous and hardy.

DAROGA RED (East)

A firm, fine-textured peach with yellow flesh and bright red skin, Daroga Red ripens after Redhaven.

DIXIERED (East/West)

A Georgia-originated peach, Dixiered has medium-sized fruit with an attractive red, slightly fuzzy skin. A semi-freestone, Dixiered has smooth yellow flesh that is non-browning. The flavor is rated as good.

This is a good-quality early-season peach, as it ripens sixteen days ahead of Redhaven.

ELBERTA (East/West)

One of America's leading peach varieties for over eighty years, the freestone Elberta and its numerous sub-varieties are superior for canning and have good shipping qualities. The fruit is large in size and is a deep golden yellow in color. The flesh has more flavor in warmer areas.

The hardy and productive tree, with large pink flowers, spreads with age.

FAIRHAVEN (East)

Introduced in 1946, this variety has been popular in the southern British Columbia interior. Productive and average in hardiness, the tree is compact. The freestone fruit is adaptable for canning and for eating fresh.

GARNET BEAUTY (East)

Originating in Ontario, Canada, the Garnet Beauty is a sub-variety of Redhaven. A semi-freestone, it has medium to large fruit of good quality with fine-textured, smooth yellow flesh. The tree is vigorous and a good choice for growers in the north.

GEORGIA BELLE (East)

An outstanding white freestone peach, the Georgia Belle is fair for freezing and poor for canning, but is excellent for eating fresh.

The flesh is white and firm, and the skin is red blushed over creamy white. The winter-hardy tree is vigorous. The fruit drops when ripe.

GOLDEN JUBILEE (East)

This peach is good for home use, for canning, and for local markets. Originally from New Jersey, the Golden Jubilee is average in quality for canning and freezing. A freestone, it has light yellow, firm, succulent flesh, with a skin that is mottled bright red. The fruit drops from the tree when ripe.

J.H. HALE (East)

The Hale, which originated in Connecticut in the early 1900s, has a reputation of being one of the finest commercial peaches. While the tree is not very vigorous, the fruit has a good flavor and keeps well.

The Hale and its several sub-varieties bear very large fruit with yellow skins that are blushed with red. The freestone flesh is yellow and free from stringiness.

HARBELLE (East)

Known for its winter hardiness, the Harbelle matures seven to nine days before Redhaven. It is a red-skinned freestone with firm, rich, yellow flesh.

HARBRITE (East)

This is a hardy freestone developed in Ontario, Canada. Good for freezing and canning, it ripens immediately after Redhaven. The skin is brilliant red over yellow; the flesh is resistant to browning.

LORING (East/West)

Excellent for fresh market sales and for processing, the Loring is a large, round fruit with a red blush over the yellow ground color on the skin and yellow, freestone flesh. It has good handling and holding qualities and is tolerant of adverse spring weather, although the tree is somewhat bud-tender.

Loring, which ripens fifteen days after Redhaven, originated in Missouri. The flowers are showy and the variety can be depended upon.

MADISON (East)

More hearty in flavor than Redhaven, the Madison ripens just before Elberta. The skin is a bright orange-yellow color with a red blush. The flesh is firm and fine-textured, like the Redhaven.

A freestone with a bright red pit, the Madison is best for local fresh market sales because of its need for careful handling.

REDHAVEN (East/West)

Originally from Michigan, the Redhaven and its numerous sub-varieties are adapted to cold climates. The standard by which all early peaches are judged, Redhaven

continued on page 32

Fairhaven

J.H. Hale

Redhaven

PROFILE:
Growing Peaches in Idaho

Gordon Saxton

An early autumn with frosty nights and cool days is not particularly welcome at the Gordon Saxton peach orchard near Caldwell, Idaho. "People stop buying peaches when it turns cold," says Gordon. "All they want from that time on are apples."

This particular bit of fruit-grower lore, along with a wealth of horticultural knowledge, has been accumulated by Saxton in some sixty years of tramping up and down orchard rows. "I started picking up brush in a Nampa apple orchard in 1921," says Saxton. "I was sixteen. Later they let me start pruning."

Not too many western orchardists have duplicated what followed. In just a few years, the Greeley, Colorado native progressed through pruning crew foreman; partner in a small Payette, Idaho apple orchard; cash renter; custom sprayer; and finally, in 1928, sole owner of forty acres southwest of Caldwell. Seventeen of the forty were in peaches, cherries, apricots, and apples. The trees were planted on some gentle slopes above Lake Lowell, a reservoir in the western Boise Valley.

Becoming his own boss did not mean instant fortune, Saxton found out. "It was really tough during the Depression years. People had little cash. Sometimes they paid for their fruit with their cream check," Gordon recalls. "I even took eggs and chickens in payment."

Peach orcharding in southern Idaho remained difficult for quite a few years, according to Saxton. "We could begin to see a little daylight in 1934. But it was not till the early forties that it appeared as if there might be a future in fruit growing."

Aided by his wife Roberta and their son Harvey, who joined the family in 1930, and by Gordon's nephew, the Saxton operation grew in size and market penetration. Sagebrush was cleared for some of the surrounding hills and neighboring orchards were acquired. By 1965 almost a thousand acres were part of the Saxton spread, with three hundred of that total in fruit trees.

Gordon, who has been called the Peach King of Idaho, believes that peaches are harder to grow than apples. "There is some truth to the saying that anyone can grow apples, but it takes a fruit grower to raise peaches," Saxton states with a chuckle.

"There is not as much margin in the raising and selling of peaches as in apples," the Caldwell farmer adds. "Plus when you sell someone a bushel basket of peaches, they have to can them or do something else with them right away. However, they can eat for a month out of a basket of apples."

When one of the authors visited the Saxtons in the early spring of 1987, field crews were busy planting new varieties of peaches and nectarines. "I have made a little money in my lifetime," states Gordon, "by having a new variety of peach or nectarine that someone else did not have. It's a gamble, but if you have it and your neighbor doesn't, then you can sell it."

The Idaho orchardist thinks nectarines may be more important on the fruit scene in the future. "The first nectarines were like eating a piece of leather," he says. "But some varieties, like the Fantasia and the Red Gold, today are just as nice to eat as a regular peach." Saxton disclosed that they have planted a sizable block of nectarines in their acreage. Despite Gordon's willingness to gamble on new varieties of fruit, he has a soft spot in his heart for Hale peaches. "I've made a pretty good living from Hale peaches," he says.

The Saxtons began selling peaches throughout the United States under their own label in the mid-thirties. Today the Saxtons are marketing their fruit with two other Caldwell area orchardists. Last year the three orchards shipped cherries and other fruit from Maine to California.

The elder Saxton proudly reveals that he is a member of the National Peach Council, a life member of the Washington Horticulture Society, and a past president of the Idaho Horticulture Society.

Gordon "sort of retired" about a decade ago, leaving the day-to-day management to his son Harvey and Harvey's wife, Phyllis. Seven granddaughters also have been helping with the Saxton family orchard operation!

However, Gordon still keeps his hand in with a local fruit stand and with a marketing operation in Nebraska. "I like my peaches to be a little riper than most sellers," states Gordon. "They have to be ripe enough for somebody to want to eat them." The local fruit stand and fast trucking direct to fruit stands in Nebraska allow Gordon to meet this goal.

There have been a lot of ups and downs in being an orchardist in the Pacific Northwest. But Saxton has no regrets, "I like this life, and I would like to go around again!"

Rosa

Suncrest

Sunhaven

has attractive red skin, and firm yellow flesh that is non-browning.

The round freestone fruit is uniform in size and is superior for fresh use and for canning. The tree is vigorous and early-bearing, hardy and highly productive. It's widely planted.

REDSKIN (East/West)

A Maryland cross between J.H. Hale and Elberta, Redskin (a freestone) is medium in size with a deep red blush covering most of the fruit. Excellent for freezing and eating fresh, the Redskin's firm flesh is non-browning. This is a good variety for shipping purposes.

RELIANCE (East)

The Reliance is a good choice for home gardens located in cold climates. Originally from New Hampshire, the dark-red-skinned peach ripens about the same time as Redhaven.

The flesh is yellow, freestone, soft and juicy, and has a good flavor. These peaches are good for fresh market sales and are delicious canned or frozen. The hardy tree produces showy flowers.

ROSA (East/West)

This freestone peach has yellow skin overlaid with a faintly streaked medium-red blush. The tree is vigorous and productive and the fruit is large.

Originating in Washington, the Rosa has very firm, slightly coarse yellow flesh that has a good flavor. Rosa is a good peach for the fresh market, for canning, and for shipping.

SUNCREST (East/West)

This large, round fruit — which ripens ten days before Elberta — is bright red in color over a yellow skin. The flesh is yellow, firm, and very flavorful. A freestone peach that originated in California, Suncrest is good for eating fresh or for canning. The tree is vigorous and hardy, and matures late.

SUNHAVEN
(East/Western Oregon)

A semi-cling, Sunhaven originated in Michigan. The sweet taste of the yellow flesh matches the bright red appearance of this fair-sized beauty. The growth habits of the tree make it large in size and in production.

VETERAN (West)

Originating in Canada in 1928, this golden-skinned peach is easy to peel but slightly soft when canned. The nearly freestone fruit is flavorful and low in acid. The tree is a vigorous grower that bears early. Veteran does well in western Washington and Oregon. It is being replaced in the British Columbia interior by Fairhaven.

Pears

AURORA (East/West)

A large, high-quality dessert pear, Aurora is very regular in shape with a bright yellow skin overlaid with russet. The flesh is smooth, melting and juicy, very flavorful, and aromatic. The Aurora keeps well in cold storage.

BARTLETT (MAX RED BARTLETT, RED SENSATION) (East/West)

Known in England as the Williams Pear, the Bartlett is a leading commercial variety. Excellent for canning and fresh use, the Bartlett has buttery-textured flesh that is distinctive and flavorful. The skin is greenish-yellow, which may be slightly blushed with red.

The strong and upright trees bear young. The Bartlett is the standard summer pear for sale at supermarkets, with the pears being picked when green for later ripening. Bartletts are the main canning pears wherever pears are grown.

BOSC (BUERRE BOSC) (East/West)

A late-ripening pear, the Bosc is dark yellow with a cinnamon-russet blush. Considered by many to be one of the finest of pears, its white flesh is juicy and tender, making it a wonderful out-of-hand eating delight.

Boscs are long-necked, large in size, and ripen best at room temperature. Originating in Belgium before 1820, the Bosc has a pleasing aroma and is a good-quality pear for market sales.

The tree is vigorous and hardy with an upright growth habit. Careful handling of the fruit is necessary.

CLAPP'S FAVORITE (RED CLAPP) (East/West)

This large, lemon-yellow pear resembles the Bartlett. Good for eating and canning, it softens quickly after picking, however, and will break down at the core if picked too late. An early-season variety, Clapp's Favorite has soft, sweet flesh and comes from a hardy tree.

COMICE (East/West)

Originating in France around 1850, this well-known dessert pear is a specialty of growers located near Medford, Oregon. A high-quality, fine-textured pear with a rosy-blushed golden yellow skin and juicy flesh, Comice develops its best flavor after a month of storage.

Bartlett

Max Red Bartlett

Comice

D'Anjou

Seckel

Many consider Comice the best winter pear. However, the tree bears erratically and is slow to bear fruit. Comice is not recommended for canning use.

D'ANJOU (RED D'ANJOU) (East/West)

A leading commercial variety in the Pacific Northwest, this large, round fruit has good keeping qualities and a superbly rich flavor. Light yellow to green in color, the D'Anjous have stocky necks and firm, mild-flavored flesh. They store well. D'Anjous ripen best after a month or two of refrigeration.

The D'Anjou pears, which originated in the Loire Valley of France and were brought to North America about 1842, are borne on a large, vigorous tree. They are not recommended for hot-summer areas.

DUCHESS (East/West)

This pear originated in France in 1820, but is a good choice for growers located in the northern United States. A greenish-yellow, very large fruit, Duchess has fine-textured flesh and a fair flavor.

MOONGLO (East/West)

Two major advantages of this variety are that the trees are resistant to the disease fire blight and they begin bearing good-sized crops when young. Originally from Maryland, the Moonglo is a medium-sized, yellow-colored fruit with white, juicy flesh.

SECKEL (East/West)

Commonly called the "Sugar Pear," the Seckel originated in New York. Not especially attractive, the small, yellowish-brown fruit nevertheless surpasses all other varieties in its dessert quality. Good eaten fresh or canned whole as a spiced preserve, the Seckel ripens late on a productive tree.

Pears, Asian
(Asian Apples) (East/West)

Asian pear

Recently popular are the Asian pears, also known as Oriental pears, Chinese pears, salad pears, and apple pears. They have a very distinctive flavor and differ from the common pear primarily in that they remain crisp and juicy when ripe and are not gritty.

Asian pear trees are fire blight resistant, and the fruit will keep in the refrigerator four to eight months without softening. Asian pears have no relationship to the apple. Some leading varieties are: Chojuro, Hosui, Kikusui, Nijisseiki, Seigyoku, Shinseiki, and Yali.

Plums

ABUNDANCE (East)

A purple-red plum, Abundance has tender, yellow flesh and is good for desserts as well as for cooking. It is recommended for northern growers.

BLUEFRE (East/West)

Another plum reported to be good for northern orchards is the Bluefre, a large blue freestone with yellow flesh. Bluefre is a late ripener, but comes from a vigorous tree that begins bearing early.

BURBANK (East/West)

The Burbank is a Japanese variety of plum that began its career in California. Large and reddish-purple in color, the Burbank is firm and meaty with a fine flavor. Burbanks are used frequently for canning and desserts. The tree, a heavy producer, does well in cold-winter areas.

DAMSON (East/West)

One of the best plums for making jam, jelly, and preserves at home, the Damson is a small blue-to-purple, semi-freestone fruit. An old variety of plum that was taken from Damascus to Italy, the Damson has exceptionally juicy, yellow flesh.

DUARTE (IMPROVED DUARTE) (East/West)

Originally from Japan, this large, heart-shaped plum that keeps well is good for eating fresh or for canning. It has a dull red skin with a dusting of silver. The deep red flesh is juicy when ripe and has a tart flavor when cooked. The trees are cold-resistant.

EARLIBLU (East/West)

This early plum comes from a hardy tree that does best when planted in northern gardens. Earliblu, which resembles the Stanley prune/plum, has tender greenish-yellow flesh.

ELEPHANT HEART (East)

A very large, heart-shaped fruit with dark purple-red skin and blood-red flesh, the Elephant Heart has a distinctive flavor and is good fresh, frozen, or canned. Developed by Luther Burbank, Elephant Heart is a freestone. The tree is vigorous and hardy.

EMBER (East)

The Ember, which originated in Minnesota, is naturally well-adapted

Burbank

Duarte

Elephant Heart

Empress

Green Gage

Italian Prune

to a cold climate. A sand cherry hybrid, Ember is a bright red, medium-to-large fruit with juicy yellow flesh.

EMPRESS (East/West)

A late-maturing, European-type plum, the Empress bears large fruit with yellow-colored, fine-textured flesh. Maturing late, the Empress is good for shipping.

GREEN GAGE (REINE CLAUDE, JEFFERSON) (East/West)

The Green Gage is a European plum dating back to 1699! The fruit is small to medium in size with amber flesh and greenish-yellow skin. Very sweet and aromatic, the Green Gage (a freestone) is ideal for canning, for preserves, and for eating fresh. The tree is hardy.

ITALIAN PRUNE (FELLENBERG, EARLY ITALIAN) (East/West)

Originating in Germany, this prune is good for eating fresh, canning, drying, and shipping. It has a purple to dark blue skin and greenish-yellow flesh that turns red when cooked. The freestone prunes are produced on a hardy tree that may overbear.

Are you confused about plums and prunes? Many people are. Just remember that a prune is a kind of plum that has more natural sugar. This extra sweetness keeps the prune from spoiling while being sun-dried.

METHLEY (East/West)

A satisfactory choice for northern growers, the Methley, which originated in South Africa, is a good commercial variety. The fruit is small to medium in size and has reddish-purple skin with red flesh and an excellent flavor.

PEACH PLUM (East/West)

The golden-yellow flesh of this round, semi-freestone plum is a delight to eat, either fresh off the tree or canned. The skin is an attractive purplish red when fully ripe.

Peach plums are an old variety, having been introduced from France to the United States in 1820.

PRESIDENT (East/West)

Originally from England, this large blue plum is slow to ripen but the trees are heavy producers and are winter-hardy. The fruit ships well, making it an outstanding commercial choice. The President is oblong in shape, and has dark yellow flesh that is good for canning and cooking.

REDHEART (East/West)

Excellent fresh, canned, or in preserves, the Redheart is a medium to large fruit with a dull green skin accented in dark red and gray. The fine-grained flesh is bright red, sweet, firm and highly aromatic. A semi-freestone, Redheart holds well on the tree and keeps well after picking. It is a California variety.

SANTA ROSA (East/West)

Developed in California by Luther Burbank, the Santa Rosa is a large, cone-shaped fruit with a deep purple skin and yellowish-pink flesh. Firm, juicy, and slightly tart, the Santa Rosa (a clingstone) is a slow-ripening plum that is good for nearly all climates.

Good fresh, canned, or in desserts, the Santa Rosa is a favorite of fruit lovers.

SATSUMA (East/West)

Another Luther Burbank creation, the semi-freestone Satsuma is small to medium in size and good for nearly all climate zones. The skin is blood red in color with the flesh being meaty and mild in flavor. It is good for desserts and preserves.

SHIRO (East/West)

A medium-round, yellow-skinned fruit with yellow flesh, the Shiro is very hardy and is considered to be among the best of the yellow plums. Good for cooking, canning, or desserts, the Shiro is widely grown in western Oregon and Washington. A very early-ripening variety, Shiro is very flavorful when allowed to mature properly.

Excellent for eating fresh, Shiro is considered a real moneymaker for local and roadside stand sales.

STANLEY (East/West)

Originally from New York, this large bluish-purple prune has greenish-yellow flesh and comes from a winter-hardy tree that produces full crops each year. Excellent for home use and processing, the Stanley, a freestone, is regarded as the most popular prune-plum. However, the fruit does have a tendency to double in hot climates.

SUPERIOR (East)

A Minnesota product, the Superior (a clingstone) is a large, conical, red-skinned fruit with firm yellow flesh that has a good flavor. Especially bred for the northern United States, Superior trees are winter-hardy.

UNDERWOOD (East)

Also originating in Minnesota, the Underwood is as winter-hardy as the Superior. A very large, red freestone with golden-yellow flesh, it is a dessert-quality plum.

YELLOW EGG (East/West)

A good choice for northwestern growers, the popular Yellow Egg plum is produced on a vigorous and hardy tree. Good for home or market sales, the Yellow Egg, a freestone, has a mild flavor.

Santa Rosa

Shiro

Raspberries

Chilcotin

Fallgold

ALASKAN VARIETIES

Varieties that have done well in the milder areas of Alaska include Boyne, Chief, Indian Summer, Latham, and Trent.

BLACK RASPBERRIES (East/West)

Black raspberries are not tolerant of mild climates. They thrive in cold and thus do poorly in western Washington, although they are planted in the Willamette Valley and elsewhere in Oregon. They bear a single crop on one-year canes. Several of the most popular black raspberry varieties are Allen, Black Hawk, Cumberland, and Munger. The latter is recommended for western Oregon, but not for western Washington.

CANBY (West)

A highly recommended red variety, the Canby is excellent for dessert uses and good for freezing. A midseason berry, the Canby ships well.

CHILCOTIN (British Columbia)

This light red berry produces semi-firm, large fruit over a long harvest season. Chilcotin is regarded as suitable for both fresh market sales and processing.

The plants are vigorous, producing large amounts of fruit.

FAIRVIEW (West)

Large, light red berries characterize the Fairview, which is recommended for western Washington gardens.

FALLGOLD (East)

The extremely sweet berries of this yellow variety are large and firm. The plant's winter hardiness makes it a good choice for northern gardens.

HILTON (East)

The Hilton, which is recommended for northern gardens, produces large, attractive red berries and is hardy.

HERITAGE (East)

This red raspberry, which is recommended for central and eastern Washington, is an ever-bearer.

MEEKER (West)

Developed by Washington State University for home gardeners, the

Meeker raspberry bears excellent-tasting, bright red fruit that freezes well for storage.

NEWBURGH (East)

The Newburgh has large, firm, red berries that ship well.

This raspberry variety is highly winter-hardy and produces its berries in midseason.

PUYALLUP (West)

This red variety, which produces soft berries, ripens late.

SUMMER (West)

Recommended for western Washington gardens, the Summer berry is red, intensely flavored, and medium to large in size.

WILLAMETTE (West)

Large, round, dark red berries that are excellent for freezing and canning typify the Willamette variety. A midseason ripener, the Willamette is widely planted in commercial plots. The berries are also good for fresh eating and in pies, jams, and jellies.

Willamette

Strawberries

ALASKAN VARIETIES

Varieties of strawberries that have been found to do well in Alaskan gardens include the Alaska Pioneer, Matared, Sitka Hybrid, and Susitna.

BENTON (West)

This variety produces king-sized, bright red strawberries that are as attractive to look at as to eat! Benton is recommended for fresh market sales rather than for shipping because of the tender berries.

The productive plants produce many runners.

HOOD (East/West)

This is a bright red, glossy, large berry that matures in midseason. It is good for eating fresh or for making jams and jellies. The Hood is the leading commercial variety in Oregon (where it originated) and is popular for home garden, as well as commercial, use.

NORTHWEST (West)

A firm, well-flavored berry, the Northwest is good for eating in desserts, in preserves, or for freezing. The Northwest berry, which originated in Washington, is resistant to viral diseases but lacks winter hardiness.

OGALLALA (East)

This variety is dark red in color, soft in texture, flavorful and good for freezing. The ever-bearing plants are cold resistant.

Benton

Olympus

Rainier

Shuksan

OLYMPUS (West)

The Olympus berries are bright red in color throughout, medium to large in size, and moderately firm in texture. A late midseason variety, the Olympus berries are good for southwestern Washington and northwestern Oregon gardens. They originated in Washington.

OZARK BEAUTY (West)

These large, sweet, bright red berries produce better in cooler than in hotter climates. They're good for freezing and dessert uses. The plants are ever-bearers.

PUGET BEAUTY (West)

Light crimson in color, highly flavored and excellent for eating fresh, in jams and jellies or for freezing, the Puget Beauty originated in Washington. These berries are mildew-resistant.

QUINAULT (East/West)

These large berries with soft flesh are an ever-bearing variety.

RAINIER (West)

Recommended for growers located west of the Cascade Mountains, the Rainier bears in late midseason. The large red berries are widely used for freezing.

SHUKSAN (East/West)

Bright red-colored, firm berries mark this midseason bearer. It is cold-resistant and recommended for northwest Washington and coastal British Columbia gardens.

TOTEM (West)

This British Columbia-developed strawberry is now the most widely planted variety in the Pacific Northwest. The solid red color of the firm yet large berries makes them a favorite for commercial processing.

Planting and Growing

Road curves through an orchard on the east shore of the Columbia, upriver from Wenatchee.

Planning to Grow Fruit

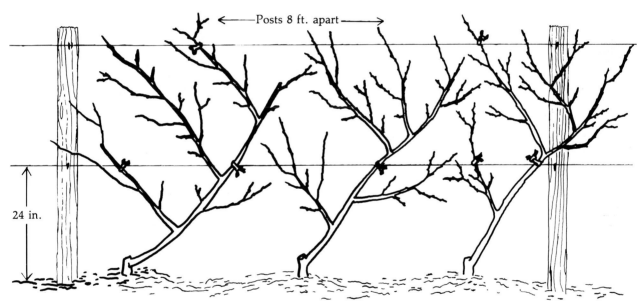

Posts 8 ft. apart

24 in.

Apple and pear trees can be trained to serve as an attractive yard divider as well as to furnish tasty fruit.

The first thing to do when starting an orchard, vineyard, or berry patch is to sit down with your coffee cup and decide what you are trying to achieve. If what you have in mind is a home installation, several cups of coffee will probably be sufficient. But if you are considering a commercial operation, better make it several pots.

In either case, it will take a year or so for your new trees and/or vines to bear fruit. This means you will invest considerable time as well as money before you see any return in the form of ripe fruit for eating or selling. Careful planning, therefore, is essential.

Location

One of the first things to consider in your planning is the location of the trees and vines. This is true whether you will be planting in your existing yard or at a commercial orchard or vineyard site.

If a home installation is your goal, some factors to keep in mind are the growth habits of the fruit varieties you are considering, the microclimates of different parts of the yard, the fertility and heaviness of the soil, and the yard's orientation to the sun and prevailing winds.

Apple tree blends into home landscape in this Dayton, Oregon garden.

Aesthetics should be considered in the placement of the trees and vines. You should especially consider how the fruit trees — with their attractive blossoms and foliage — will blend with or accent your landscaping. Fruit trees and grape vines trained in an espalier fashion work well as a yard divider, for instance.

Consider as well your vegetable garden plans. Vegetables can be grown in close proximity to your fruit trees and vines, thus giving you maximum returns from your available yard space.

If you are planning to go commercial, the location factor is even more important. In fact, it may be the most critical aspect.

Closely involved in the question of location is the kind of orchard or vineyard you are considering. Are you thinking big or small? Is the orchard or vineyard going to be your major business, or just a sideline? Are you thinking U-Pick or owner-do-it-all? Will you need seasonal laborers? Are they readily available?

Will you sell your fruit yourself in a fruit stand? Or are you going to market it through a nearby grocery store or supermarket chain?

Take the High Ground

If you are planning to purchase land on which to plant a commercial orchard, and you have a choice, most growers recommend staying away from the bottom lands of a valley. Choose instead the gentle slopes on the sides of the valley where, come springtime, cold winter air will drain downslope, away from your tender buds and blossoms.

If your proposed orchard will be on flat land, avoid acreage with low pockets where cold air can accumulate. If cold air pockets are unavoidable, you will need to protect the blossoms with wind machines or water sprinkling.

Home growers also need to think about air drainage. If cold air stacks up against the side of your house, garage, or thick hedge, the night-time temperature may be critical in such locations.

The converse may be true during the sunlight hours. The reflection of sunlight from a building wall may raise the temperature of a nearby spot to either harmful or beneficial levels. Some careful measurements with a good thermometer could show that your yard may have several microclimates that you can take advantage of or avoid as needed.

Sun, Wind, and Water

Fruit trees, grape vines and berry plants, like most vegetation, need a lot of sunlight to energize their photosynthetic activities. The total amount of sunlight will depend upon your geographic location and weather patterns. However, orienting the home or commercial orchard/vineyard so that the sun has a full daily sweep across the rows will be a definite plus.

Sites subject to strong winds will need the addition of tall trees or some type of fence that deflects the wind up and away from direct passage through the trees and vines. Winds low in humidity can take a significant amount of moisture away from the plants.

A good supply of water is essential, unless you are located in a humid coastal region. Fruit trees and grape vines require consistent watering to maintain their vigor and to produce abundant crops. If you live in an established town, your municipal water supply probably will be adequate for a home installation. However, the question of water supply becomes much more complex for the would-be commercial orchardist/viticulturist desiring to locate in a rural area. Depth to a producing aquifer, pump and electricity costs, irrigation canal water availability, water rights, governmental agency regulations — all these must be taken into account.

Soil

Not all earth is Good Earth, especially for the growing of fruit trees and grapes. You should have the soil of your yard or prospective commercial site analyzed for its potential fertility, acidity, and mineral balance. It is far easier and cheaper to correct any deficiencies before planting rather than afterward.

Espalier treatment provides an attractive focal point and allows plants to take advantage of reflected heat.

Apple blossoms are visual treat.

Drainage is important too. Pear trees will tolerate a soggy soil but hardly any other kind of fruit tree will tolerate its roots sitting in saturated soil. A little work with a shovel or posthole digger will tell you if you have the undesirable condition called "hardpan" underlying the top few inches of soil. The hardpan generally will stop the downward filtration of water, so that the soil at the root level will be saturated. Hardpan also will prevent the roots from working their way downward to establish a healthy root system.

Choosing Varieties

Next, you need to think about what fruit varieties you plan to grow in the home or commercial orchard/vineyard. This step calls for extensive study of nursery catalogs as well as other publications and books in your local library, bookstore, or county extension office. Choose varieties that mature well in your area. Other factors to keep in mind are vigor of the variety, the fruit's flavor, ripening dates, the initial bearing age, disease resistance, and suitability for eating fresh, for cooking, and for baking.

Visits to your local nurseries and to other orchards or vineyards in your area are a must. Your local agricultural cooperative extension agent can be an invaluable source of information about how different varieties have fared in your area. Additional sources of help are the agriculture departments of the various states, horticulture associations, fruit commissions, marketing agreement groups, and such organizations as the Cherry Institute.

The fruit variety chosen will determine how many trees you will be able to plant in your available space. As discussed in the chapter on the fruit industry, the trend in commercial orchards is toward dense plantings that make it possible to be more efficient in the use of labor and the application of chemicals. A small tree is easier to prune, spray, and harvest. It will not need as much spray to cover its leaves nor as much fertilizer to spark its growth.

When choosing the varieties of trees you wish to grow, especially for a backyard orchard, keep in mind the need for cross-pollination as well as the different times at which the fruit will become ripe. Some varieties, especially apples, are not self-pollinating. They need pollen from another variety to fertilize the blossoms. Some kinds of crabapples are pollinators with extra benefits thrown in — you get the attractiveness of these handsome trees in bloom.

If a commercial orchard is your goal, you must consider the availability of bees to do what they do best — carry pollen from one blossom and tree to another. Are there commercial beekeepers in your area

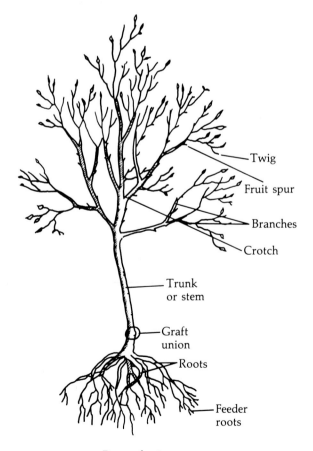

Parts of a tree

and will they be interested in placing their beehives in your orchard at the proper time?

The maturation date of fruit can be critical to the success of an orchard business. First on the market generally means a better price, although flavor and keeping qualities are also significant. The home orchardist also needs to think about when his or her fruit trees will bear their fruit. It probably will make more sense to have several varieties, space permitting, that mature at different dates. That way, the fruit can be picked and enjoyed at peak flavor.

The Man-Made Dwarfs

If you want to plant a fruit tree that in several years will provide you with shade in your yard, then you want one of the older standard-sized trees. The new modern dwarfs that nurserymen have developed are too small and too low! The World War II song that urges a young lady not to sit "under the apple tree with anyone else but me" definitely would not apply to today's genetic dwarfs.

The branches of a standard apple tree will cover a circle about forty feet in diameter when fully grown. The dwarfs, however, may spread out three to ten feet. But this small size does not mean drastically smaller amounts of fruit. The fruit production of these dwarfs will amaze you. The dwarfs also bear earlier than the standard-size varieties.

Nurserymen achieve a maximum dwarfing effect by grafting a dwarf form of a fruit tree onto rootstock that also has dwarfing characteristics. In addition, the branches and roots can be pruned so as to cut the production and flow of nutrients from the leaves and the roots. Still another method is to girdle or score the tree trunk to interrupt the nutrient movement. A narrow circle of bark may be removed from the trunk or a cut made all the way around the trunk. The tree eventually will repair this intentional damage, so the girdling or scoring will have to be repeated.

This dwarf apple tree bears plenty of fruit.

PROFILE:

"Goat Patch" Orchards — Canada's Okanagan

Anne and Jake Van Westen

The local beekeeper calls them "goat patches." Nevertheless, Jake Van Westen, a Naramata, British Columbia fruit grower, still likes the orchards of his area. "Yes, they are small and some are sloped. But we seldom have frost pockets because the air drainage is good," he says. "It is true that most growers have a number of holdings. But unless they are widely scattered, that is not too much of a problem."

Perhaps Tim Smith, who sells honey and rents hives out to orchardists such as Van Westen, sees it differently. He has to transport hives, full of temperamental bees, up and down twisting roads and across the steep angles of the orchards. Once in a while the hives are spilled and overturned. And as Smith can tell you, bees have a very effective way of showing their displeasure to nearby humans!

There is no question, however, that Van Westen and his wife, Anne, live in a lovely region. They have a superb view from their home, sited on a bench several miles south of the tiny community of Naramata. A well-trimmed orchard slopes down from their modern house to the Naramata-Penticton road. Beyond the road, still other orchards descend to the blue water of Okanagan Lake. Beyond the lake, some really steep hills and forested ridges fill their western horizon.

The limited size of many of the area's orchards — the Van Westens have separate holdings of ten, eight, twenty-seven, and fifteen acres — is partly a result of the soil and the topography of the bench-lands above the lake. Numerous large gullies have been eroded down through thick deposits of the whitish-tan silt making up most of the benches. Ancient volcanic flows also have cut up the land-scape. The end result is the small fields.

"There is no way we can match the size of some of the new orchard plantings in the United States," says Van Westen. "We simply do not have enough flat land with the proper soil and climate."

However, the Dutch-born fruit grower is not ready to concede the fruit "war" to his orchard brethren south of the border. "There will be difficulties down the road if all those new plantings in Washington state come into production — and stay in production," Van Westen states. "However, I think there always will be a future for orchardists who produce fruit of good quality."

The Naramata resident — who is highly regarded by the Okanagan region's horticulturists — believes one solution to the looming crop surpluses is market expansion. "Some improvement has taken place but much more needs to be done."

According to Van Westen, most Okanagan fruit is sold in western Canada. "Some is shipped to Ontario and Quebec as well as to major U.S. cities by B.C. Fruits, our marketing agency," he states. The province's fruit growers contract on a five-year basis with B.C. Fruits to sell their output.

A small amount of Okanagan fruit is sent to Asia and Europe, according to Van Westen. "The jumbo planes now used on the cross-ocean flights have helped to bring the freight costs down," he says.

Like those of many of their neighbors, the Van Westens' orchard is a family operation. Their son, Jake, Jr., works in the orchard and Anne's father helps out. They do have to add help when their cherries and apples ripen.

"All of our apples, mostly Golden and Red Delicious, ripen over a one-month period," states Van Westen. "The same is true of the cherries. It makes for busy times." The Naramatans grow Vans, Lamberts, Stellas, and Sams. They are now starting to grow some of the Lapin variety, expected to have good resistance to cracking.

The temporary help engaged by the Van Westens is mostly high-school-age youth from Canada's eastern provinces. "Many of them are Quebeckers," says Anne. "They like to come west to earn some extra money. The weather is good during the picking season, which allows them to sleep outside."

Jake began his training as a fruit grower early in life, by attending a four-year horticultural school in his native Holland. "We had classes four days a week for the first two years," he states. "Then we had classes three days of the week. This allowed us to obtain practical experience in the vegetable fields and orchards."

In 1954, Van Westen emigrated to Canada, ending up in Vancouver, British Columbia after a sea and train voyage that took several weeks. "I did not know the language," states Jake, "but I thought the snow-covered mountains really were something." Following a Naramata orchardist's request for help to a British Columbia employment agency, Van Westen and a fellow Dutch emigrant were sent to the Okanagan area.

In the years that followed, Jake worked in Naramata orchards, then tried renting some acreage

Bee hives are a common — and necessary — part of the orchard scene.

farther south in the Oliver area. "It was a frost pocket," relates Jake. "We were frozen out three years in a row." Finally the Van Westens — Jake had married Anne, also a native of Holland, in 1959 — began acquiring acreage of their own around Naramata.

"I thought the Okanagan Lake area was really beautiful when I first saw it in '54," states Jake. It was something to see — the lake, the fruit orchards, the distant hills.

"I haven't changed my mind."

Planting Procedures

Once you have completed your planning — or exhausted your coffee supply — the next step is to prepare your orchard, vineyard, or berry patch site for planting. In the case of a commercial facility, there are three possibilities. One is planting in virgin soil, a second is replanting of new varieties in an old site, and the third is the replanting of the same varieties where trees have died.

Most experienced growers will opt for the first choice, if possible. Virgin ground will not have been exhausted of the necessary growth minerals; the ground will not be compacted from the use of tractors; and the soil will not be harboring disease organisms left over from previous plantings.

If your new trees or vines will be planted on a totally new site, remember to smooth the contours of the ground as much as possible. This leveling will aid in air drainage and keep water from ponding in low spots.

Many nursery experts recommend cultivating soil of a prospective orchard or vineyard for one to three years before planting any trees or vines. Tilling the soil, growing green manure crops and turning them under, and fumigating the soil if necessary can be done during this time.

It is possible to use the site of a previous orchard or vineyard by following a careful program of soil tilling, fertilization, and disease eradication. It is strongly recommended that any leftover tree roots, bark, or other debris be removed from the soil. Fumigation with an appropriate chemical is a good idea to kill off any disease organisms left behind in the ground.

1) Dig hole twice the size of root system

2) Mix native soil with compost, peat moss, or well-rotted manure

3) Spread roots over cone of soil

4) Put first stake on windy side of tree

5) Water thoroughly

Follow a few simple steps when planting a tree. ▶

Fall or Spring Planting?

When you should plant your new fruit trees will depend upon where you live in the Pacific Northwest. If there is danger of your soil freezing in the winter months, you should plant in the spring. However, if your soil does not freeze, putting young trees in the ground during the fall months will allow them to become established and ready to grow, come the warm days of spring. Fall plantings help the root system to become established before the leaves and branches start their growth.

Bare Root or Balled?

Fruit trees and vines usually are shipped from nurseries in a bare root condition; that is, there is no soil around the roots. The nursery will pack some wood shavings, peat moss, or other waterholding medium around the roots to keep them from drying out during shipment.

Once the plants have been delivered to you, this moist condition must be maintained. You can add water to the shavings or, better yet, put the trees or vines on their sides in a shallow trench, cover the roots with soil, and keep them moist. Experienced gardeners call this process "heeling-in."

If you wait to purchase your new plants from a nursery until mid-spring, you probably will get them with the rootball encased in burlap or in a plastic or metal container. Once a tree begins to sprout leaf and branch buds, the nursery worker must add soil around the roots. Needless to say, more will be charged for trees sold to you in this condition.

Young trees balled in burlap or in a container also should be kept from drying out. Sprinkle the outside of the burlap or add water to the container.

When planting burlapped trees, place the entire burlap ball in the hole. The burlap will decay after it has been in the soil for several weeks. Trees and vines received in containers must have the containers removed before they are planted. Do try to keep as much as possible of the soil around the roots during the planting process.

Two-Dollar Holes Not Recommended

The good advice — do not plant a ten-dollar shrub in a two-dollar hole — applies to fruit trees and vines as well as to ornamental plants. An adequately sized hole should be dug. The sides of the hole should be

Mulching these blueberries conserves moisture.

straight up and down, not slanting. The soil should be loosened in the bottom of the hole. Some organic material, such as compost, peat moss, or shredded bark, can be added. If you add bark, you will eventually need to add fertilizer around the plant to make up for nitrogen losses from the soil when the bark decays.

Do not mix fertilizer in the planting hole or in the soil that you place around the roots. Doing so runs the risk of burning the roots with the fertilizer.

The soil placed back in the planting hole should be the soil that you took out. Do not take it from elsewhere in the yard, as a different soil can affect the growth of the roots.

Some careful pruning of the roots should be done when the tree or vine is planted. Any damaged roots should be cut away and some pruning done to balance the top growth (which will emerge in spring) with the root structure in the ground. Some nurseries recommend cutting away one-third to one-fourth of the root structure when a tree is planted. If the young tree has no branches, cut the top off so that 30 to 36 inches of trunk remain. If branches are present, cut away any weak or broken ones. Cut one quarter of the length from the remaining ones. Cut the header down to about a foot and a half above the top branch.

Once the young tree or vine is in the ground and the planting hole backfilled with the original soil, you should tamp down the soil to eliminate any air pockets. The round end of your shovel or your shoe heel makes a good tamping device.

Staking a tree prevents damage from wind.

Bud Union Level

Make sure that the bud union — where the nursery grafted the fruiting part of the tree onto the rootstock — remains above the soil. Otherwise you may have new branches emerging from below the union. The end result will not be the fruit variety you thought you were buying!

Staking, Watering, Mulching

Most small fruit trees and vines will not require any initial staking to prevent damage from wind. However, a large tree, such as one that came from the nursery in a container, probably will need staking. Add two stakes, one on each side of the tree. Tie the tree to the stakes, using some type of material that will not damage the bark of the tree. Definitely do not use wire!

Both bare-root and container trees and vines should be planted approximately at or above the same level as they were in the ground at the nursery. The nursery planting level shows as a change in bark color.

When watering-in new trees and vines, use soil to build a small basin around the tree. Make it slightly larger than the root ball. Fill the basin with water, letting it soak into the ground. If the tree or vine sinks at this stage because of the watering, slide a shovel underneath the root ball and raise it slightly. Add more soil to keep the root ball at the proper height.

Once the new trees or vines are in the ground, you need to develop a regular watering schedule — unless you are located in a coastal region. As with ornamental

trees and shrubs, a regular schedule of watering deeply and infrequently is better than lightly sprinkling the soil around the plant. An easy way to check the soil moisture is by poking a small rod into the soil. The distance the rod penetrates will demonstrate the moisture depth.

You might consider the use of drip or micro-jet methods of irrigation for your trees and vines. With the drip method, the water simply drips or oozes from a network of plastic pipes and emitters directly to the root zone.

Another effective moisture-conserving measure is to place a mulch of bark, leaves, or straw around your trees or vines. This mulch will help conserve moisture as well as help control weed growth. However, be sure that the mulch does not touch the trunk. Keep it several inches away, otherwise mice can hide in it while munching on the bark!

Watering should be discontinued in the early fall so that the trees and vines will go into their dormant stage. Otherwise you run the risk of winter kills of new growth encouraged by late watering.

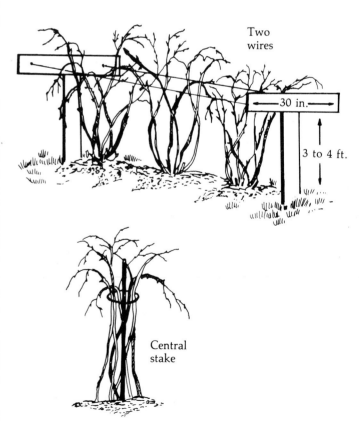

Raspberry brambles should be gathered around a central stake or grouped between two wires.

Pruning Guidelines

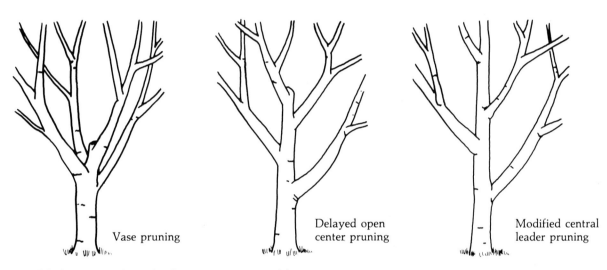

Simplified drawings show the three main systems of fruit tree pruning.

Vase pruning

Delayed open
center pruning

Modified central
leader pruning

Pruning your fruit trees, grapes, and berries is the vital step that can make all the difference between straggly, unattractive trees, vines, and canes and neat specimens that will be a delight to your eye as well as to your larder — or to your pocketbook if you are planning to be a commercial grower. The "natural look" may be appropriate for some flower gardens or for fashion styles, but not for an orchard or vineyard. What you are after is fruit production, not nature's survival of the fittest.

Pruning requirements will vary from apples to peaches to pears. What we can give you here are just some general rules that will help you get started. You should supplement this information with specific publications from your local county extension service, attend some pruning clinics if offered, and "take a leaf" from a neighbor whose trees, vines, and berries are doing well.

Begin by acquiring some good-quality tools. Long and short-handled pruning shears, coarse-toothed pruning saws (a bow saw for large limbs and a folding saw for the smaller branches), and a sturdy ladder with an extendable third leg — all will come in handy. A good ladder is especially important as you will not enjoy your fruit as much with a leg or arm in a cast!

Neither Timid Nor Overbold Be

Home growers generally take one of two approaches to pruning their trees and vines. Some are intimidated by the task and make a few timid cuts that are more cosmetic than effective. Others start hacking away with great zeal at whatever limb or branch is at hand. What is needed is an approach somewhere between these two extremes.

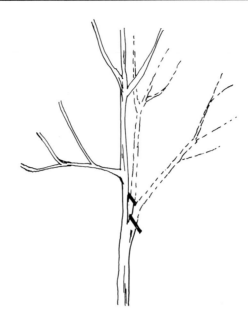

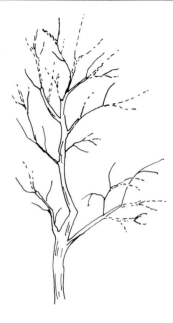

Too many fruit spurs are too much of a good thing. Thin them out and the result will be larger fruit.

Narrow-angled branches should be removed before the tree matures. Otherwise they may later split away, ruining the entire tree.

Once a fruit tree matures, keep it producing by thinning out the shoots toward the ends of the branches.

Before
thinning

After
thinning

◄Although it will seem to go against the beginning orchardist's instincts, a heavy crop of small fruit early in the growing season should be drastically thinned.

It helps to have a plan in mind for your particular orchard, vineyard, or garden. You need to know what the growth habits of your varieties are in relation to the size of your growing area. Then take some time to study each tree or vine before you begin. Try to visualize how it will appear next year, after the new growth has emerged following the pruning season.

There are some obvious things to do with your shears and saws. Cut off any dead or diseased wood, eliminate branches that are rubbing against each other, and get rid of branches that form less than 45-degree angles with the trunk. These may later split away from the trunk, ruining the entire tree in the process. (This statement does not apply to pears, as they have a naturally upright growth habit.)

Do not leave any stubs of branches; cut them off flush with the trunk. Also remove any root suckers.

Let the Air In

Next, and not so obvious, is to prune so as to limit the growth of the specimen while at the same time

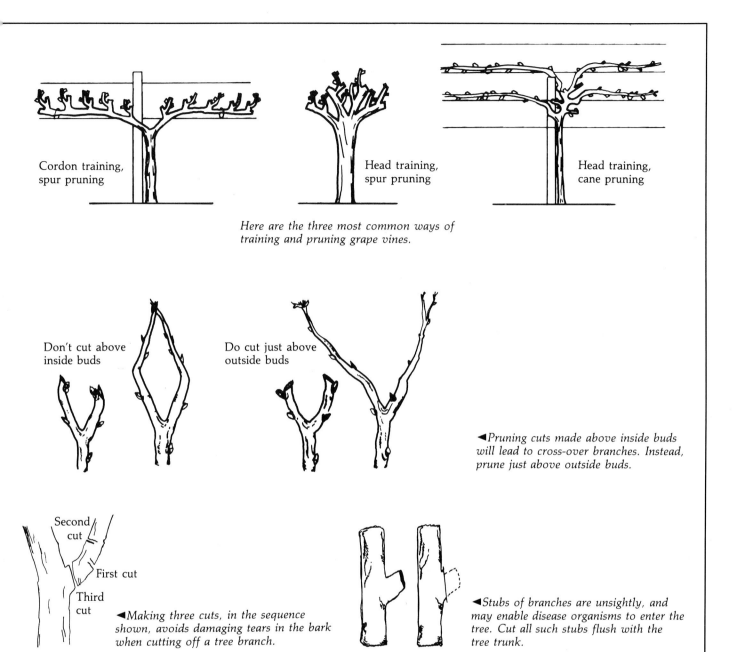

Cordon training, spur pruning

Head training, spur pruning

Head training, cane pruning

Here are the three most common ways of training and pruning grape vines.

Don't cut above inside buds

Do cut just above outside buds

◄*Pruning cuts made above inside buds will lead to cross-over branches. Instead, prune just above outside buds.*

Second cut

First cut

Third cut

◄*Making three cuts, in the sequence shown, avoids damaging tears in the bark when cutting off a tree branch.*

◄*Stubs of branches are unsightly, and may enable disease organisms to enter the tree. Cut all such stubs flush with the tree trunk.*

maintaining healthy fruiting wood that will be open to sunlight and air movement throughout the tree or vine. These goals generally call for judicious removal of the older, heavier limbs in the top of fruit trees.

As a general rule, the younger the tree or vine, the more lightly it should be pruned. Many gardeners seem to take the opposite approach, apparently thinking that once the specimen has attained its full size, it should be left alone. However, this is not recommended.

After a newly planted tree or vine has been clipped to balance its top growth potential with its root system,

it usually should be left alone until it begins to blossom and bear fruit.

Subsequent pruning should be done during the time the tree or vine is dormant. Doing so lessens the possibility of freeze damage. However, pruning may be done during the summer months to reduce the vigor of dwarf and semi-dwarf varieties.

Cord, weights and short boards also can be used to train the growing branches of fruit trees in a desired direction. The goal is to expose the fruiting wood to the sunlight so as to maximize the fruit growth.

PROFILE:
Start with Old Apple Trees . . .

Perkins Orchard in Skagit Valley

Three old apple trees started it all.
"When we retired here in 1968, there were these run-down, overgrown apple trees on the place," states Tom Perkins. "They were a mess.

"But I didn't know anything about apples, except to eat them. So I took a home orchard short course at the local community college, taught by Bob Norton, director of the agriculture research station at Mount Vernon. He taught us how to plant, prune, and graft. He also introduced us to the new apple varieties growing at the research station.

"From that time on we have continued to graft and plant. Now we have sixteen thousand trees on twenty-six acres and another twenty acres to plant."

What Tom and his wife Sue have on their relatively small acreage is one of the Pacific Northwest's most unusual orchards. Tucked into the lush

Grafting an apple tree

bottomland of the Skagit River Valley near the Washington state community of Sedro Woolley, their sixteen thousand apple trees are a magnet for other orchardists. These growers want to see the many varieties grown by the Perkinses as well as the different horticultural methods used by them.

Most people think of apple growing in the Pacific Northwest in connection with the Canadian Okanagan region, the Wenatchee and Yakima valleys of Washington, and the Hood River area of Oregon. The cool coastal climate of the Skagit Valley is not considered to be apple country. However, Tom and Sue have been proving them wrong with their Sedro Woolley operation. "Sometimes I get as much or more tree growth as they will in eastern Washington," says Tom. "And I do it without any irrigation or fertilizer, once the trees have been planted in their permanent location." The grower adds that his orchard requires less spraying to control diseases and pests than the orchards on the eastern side of the Cascade Mountains. Scale and mildew are the major problems in this area. (As noted in the section on pest control, there are fewer insects on the west side of the Cascades.)

Tom and Sue are quick to point out that their remoteness from other orchardists has helped them be commercially successful. There are several other relatively small orchards in the Mount Vernon-Burlington area, but nothing approaching the density of the orchards in eastern Washington and the Canadian Okanagan.

Growing many different varieties and doing most of the orchard work themselves are two additional reasons why the Perkinses sell their entire crop most years. "We are unique in that we grow so many varieties," Tom, a native West Virginian, states. "We try to cover the entire length of the fresh fruit season with about a dozen varieties."

"We sell about half our apple crop to people who drive out here," Sue says. "Most are repeat customers." The other half of their apple crop goes to a couple of grocery store chains.

The Perkinses found out early in their orcharding career that their western Washington location meant that some of the well-known apple varieties would turn out differently. "We do not get enough heat units here for the Delicious or the Granny Smith to mature," Tom states. He emphasizes that different growing locations will make the same apple variety look and taste different. "Compare the Spartan variety, for example," Tom explains. "You probably would not recognize a Spartan grown here as compared to Spartan coming from the Yakima Valley."

The moral of the Perkinses story might be to approach with caution the pruning of any old trees already on one's property — it could become a long-term habit!

Tom and Sue Perkins

Pests and Diseases

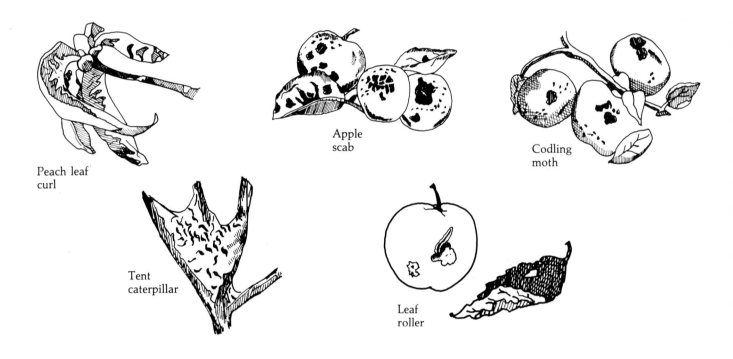

Peach leaf curl

Apple scab

Codling moth

Tent caterpillar

Leaf roller

A multitude of hungry insects, fungi, viruses, and bacteria will be eager to fatten themselves on your new fruit crop. A few of the worst offenders are sketched here. Most successful growers follow an established routine of preventative sprays coupled with clean orchards and gardens. Detailed information should be obtained from your state or province agricultural service, local nurseries, or neighbors whose trees and vines are doing well.

Many persons, particularly those interested in their gardens, probably would prefer not to use chemical sprays on their fruit trees. However, it is a well-known fact that there is a mighty host of insects, fungi, spores, and bacteria lurking in most gardens, which like to "dine" on tasty fruit, as well as on the trees and vines. Experienced orchardists and viticulturists are convinced the only way to have unblemished fruit is by following a careful and sustained program of spraying. The measures required will depend to some extent on the orchard or vineyard location. In coastal areas, diseases are the major problem. For some of these afflictions there is no alternative to spraying. But in semi-arid regions, insects will be the grower's big headache. An integrated approach to insect control may make it possible to reduce pesticides.

Some reduction in spraying can be achieved by maintaining the orchard and vineyard area in a super-clean state. All dropped leaves and fruit, as well as any diseased twigs or limbs, should be completely removed from the area — or burned if local regulations permit.

Keeping the individual trees and vines as healthy as possible through proper pruning and fertilization also will cut the need for chemical sprays. Healthy plants are able to keep many diseases at bay. Just as in the animal world, predators have greater success attacking the weaker specimens.

Growers should strive to be good neighbors by controlling the pests and diseases in their own plot or acreage. Insects and wind-borne diseases do not respect property lines. Each grower has a responsibility to neighbors to keep trees disease and pest-free.

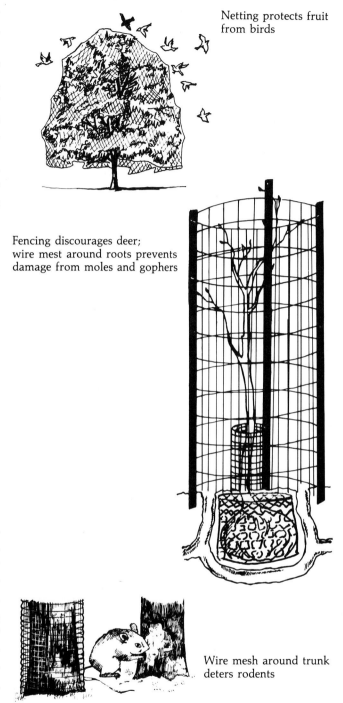

Netting protects fruit from birds

Fencing discourages deer; wire mest around roots prevents damage from moles and gophers

Use great care when spraying an orchard.

Detailed information about the various pests and diseases that may affect fruit trees and vines is beyond the scope of this publication. Consult with your state or province's local agricultural extension service representatives for publications and short courses that will give you valuable guidance. Chemical manufacturers also publish helpful pamphlets, and the labels on their products are instructive. If you are considering a commercial orchard or vineyard, contact field representatives employed by the manufacturers and distributors.

It is worth stressing that all instructions on the chemical containers should be followed carefully. Twice the dosage is not twice as good!

The use of protective clothing, gloves, and respirators (where advised) should be observed as well as the proper disposal of pesticide containers. Remember too that there are many beneficial insects that eat other fruit pests. Indiscriminate spraying can kill them.

Wire mesh around trunk deters rodents

Four-footed pests, as well as some with wings, will appreciate your furnishing them with gourmet food. Moles and gophers will go for the roots, mice and rabbits think bark and leaves are tasty, deer are attracted to the leaves and fruit, while birds believe cherries and grapes are vital to their dietary intake. Some repellents are available, but physical barriers such as wire mesh, fencing, and nets are more effective.

PROFILE:

The Perils of Fruit-Growing —
Montana's Beautiful Bitterroot Valley

Charles Swanson

In the early 1900s the pinkish-white blossoms and graceful green branches of Red McIntosh apple trees stretched for fifty miles down the narrow but beautiful Bitterroot River Valley of western Montana. Thousands of acres of the fragrant blooms greeted spring each year as the snows began to melt from the nine- to ten-thousand-foot peaks rimming the valley's western border.

Today, St. Joseph, Canyon, and Trapper peaks are still there, but one 25-acre commercial orchard is all you will find in the Bitterroot Valley. This one remaining enclave of apple trees belongs to Charles and Julie Swanson of Corvallis, Montana.

Between the two levels of orchard development — the thousands of acres of bearing fruit trees during the early 1900s and today's small but brave patch of the Swansons — lies a lot of western history. It is replete with stories of land development, community boosterism, immigrants, insect infestations, orchard failures — and now, a potential rebirth of Montana's apple industry.

What happened to the Swansons illustrates this history. Charles's grandfather, also named Charles, came to the United States from Sweden in 1902. He began his life in America as a millwright in a piano factory in Illinois. Later, wanting to get away from the dust of the factory, he moved west to Montana, where the Rocky Mountains catch the incoming westerlies from the Pacific Ocean. The higher terrain forces the moisture-laden air upward, where it precipitates out as snow in the cold months and as rain during the spring and autumn.

Such weather and terrain, if not too extreme, can be good conditions for raising the hardier fruits, such as apples. Shortly after the turn of the century, some businessmen thought they had found such a place in the Bitterroot Valley, just south of where the river merges with the Clark Fork River.

The long and relatively flat valley, with its abundant water supply available from the snow fields of nearby peaks, attracted the attention of early stockmen, farmers, and land developers. By 1910 a number of irrigation companies had been organized, with such seductive names as the Sunset Orchard Land Company, the Bitterroot Stock Farm, the Bitterroot Plantation Company, and the Bitterroot Valley Irrigation Company.

Modest promises were not the stock in trade of these organizations. "Unprecedented opportunities are offered in abundance. The former isola-

Fruit trees in winter

tion of this valley has left to this generation a flood of chances for home, independence, and fortune . . . the greatest irrigated fruit project in the world has just completed its irrigation system and now offers at comparatively low prices those opportunities which were years ago commonly to be found in Yakima, Hood River, and Wenatchee valleys," were some of the advertising claims.

Prospective land buyers and investors were greeted warmly by the Bitterroot land developers. They were housed in an elaborate hotel (designed by Frank Lloyd Wright), constructed for the purpose of wining and dining them.

These advertising pitches of the irrigation companies must have been effective, for within a decade, Bitterroot apples were a force on the national market. One of the selling points of the Bitterroot orchardists was the freedom of the valley and its apples from many of the diseases and fruit pests that were afflicting other regions. "You can eat a Bitterroot apple in the dark!" was the slogan they adopted, thus attesting to their fruit's freedom from worms. Alas, this immunity was not to last.

When Charles's grandfather came to the Corvallis area of the Bitterroot Valley in 1909, he acquired ten acres of benchland along the eastern side of the valley. According to Charles, his grandfather planted five hundred apple trees, mostly Red McIntosh, on those ten acres. (Some of those original trees are there today. They are showing their age, but they are still producing.) The rest of the Swanson acreage was put into different crops and used for stock raising.

In the 1930s the promotional plans of the Bitterroot irrigation developers turned wormy. The fruit pests and diseases that had once left the valley alone decided to stay for breakfast, lunch, and dinner. "The insects came," says Charles, "and many of the valley's small orchardists could not cope with them. They lacked enough capital to invest in the needed spraying equipment and chemicals."

Some of the growers also had resorted to the practice of "top dressing" their apple boxes with the best specimens on the top, according to Charles. "This was a shoddy practice and hurt the reputa-

tion of the valley's fruit."

These factors, coupled with stiff competition from the more established apple regions of Oregon and Washington, led to a slow decline of the Bitterroot orchards. By the 1960s only remnants were left. The Swansons' ten acres of McIntosh remained, however. "My father, Carl, did not add to the plantings," relates Charles, "but he kept the orchard alive."

When Charles finished a tour of duty in the U.S. Army Veterinary Corps in 1974, he calculated that it was time to give Bitterroot orcharding another shot. "We planted 2,500 trees, some in '74 and some in '78," he states. Ninety percent of the new plantings are of the Red McIntosh variety. However, he and his wife Julie also have Alexander, Red and Golden Delicious, Paulared, Gravenstein, Viking, Spartan, and Empire trees.

In the fall of 1986, Charles was optimistic about the future of the Swanson family orchard operation. "The severe frosts of the last two springs cut my total crop way down," he said. "But most years we don't break records for cold weather during blossomtime as we did in 1984 and 1985."

The Montana orchardist reports that the Bitterroot Valley has from 140 to 145 days as a growing period. "We usually have a mild frost during the first week in September. Then in mid-October, we get a killer."

The Swansons sell much of their fruit to people who drive out over winding roads in the autumn to the Swanson orchard near Corvallis. "I like to sell the apples retail. Many people reserved boxes of apples this year as early as June."

In years when the weather permits the production of normal crops, the Swansons also sell the red-colored, tasty fruit to large grocery store chains in the state. "I can sell everything in Montana that I can grow here," states Charles. "The market in Montana is very good."

So perhaps things are changing in one of the Pacific Northwest's handsomest valleys. Maybe once again there will be mile upon mile of pinkish-white blossoms come springtime, blooming below the snow-capped heights of St. Joseph and Trapper.

The Fruit Industry

A deserted pioneer orchard makes a picturesque scene.

Early History

Fruit growing has been part of the Pacific Northwest scene ever since the very early pioneer days. No sooner had the Hudson's Bay Company set up shop at Fort Vancouver — across the Columbia River from where Portland, Oregon now sprawls — than fruit trees were planted. Employees of the famed fur-trading firm tucked apple seeds, brought from England, into the riverbank soil.

Less than a decade later, when Oregon Trail wagon trains began to reach Fort Vancouver, the fruit made available by the fort's generous factor, John McLoughlin, gladdened many a weary traveler's heart and palate. In fact, one such pioneer was so taken with the Oregon Country fruit that he reportedly loaded his wagon with apples and set off southward to sell them in California.

The success of this early merchandising effort is not known. However, before too many more years passed, the transporting of fruit to the Golden State would have been similar to "carrying coals to Newcastle." Commercial production of apples in California, particularly in the Pajaro Valley around Watsonville, began in the mid-1800s, with shipments being made to distant United States cities around the end of the century. This was well in advance of any Pacific Northwest efforts.

Making Up for Lost Time

Nevertheless, when the fruit growers of the Pacific Northwest finally got in motion, they made up for their tardiness. By 1917, Washington state, for example, was the leading U.S. producer of apples. New York, which for years had been the nation's leader in commercial sales, was relegated to second place. Additional contributions to the Pacific Northwest's fruit production total came from Oregon, Idaho, Montana, and British Columbia.

Most of this growth in Pacific Northwest commercial orchards had taken place shortly after 1900. At first, the growers of the more established fruit-raising areas of western New York, the Piedmont region of Virginia, and the Shenandoah and Cumberland valleys of the eastern United States might not have taken the western orchardists seriously. These would-be apple raisers of the West had good soil, good climate, and plentiful irrigation water in the region's mountainous snowpacks. But where did these Washington, Oregon, Idaho, Montana, and British Columbia upstarts think they would sell their fruit? As the easterners knew, the markets were in the large cities — New York, Chicago, Boston, Washington, D.C. — and in Europe.

Considerable capital was going to be needed to clear the sagebrush and pine trees from the western mountain terraces and foothills, as well as to bring the snowmelt water to the dry soils. Before long, however, the New York and Virginia orchardists began to view the new growers of Washington's Wenatchee and Yakima valleys, and Oregon's Hood and Rogue River regions, with respect. And well they might. J.C. Folger, then assistant secretary of the International Apple Shippers Association, and S.M. Thomson, a fruit crop specialist with the U.S. Department of Agriculture, described in 1921 what happened:

> The productiveness of such valleys as the Yakima and Wenatchee in Washington was phenomenal. Trees were young and free from disease, the yields on bearing trees were unusual, and the returns to the acre were far greater than had been thought possible from any commercial orchard.

Political Pressure Unavailing

Once the eastern fruit industry became aware of the dimensions of this western threat, they put on the political and economic pressure. Keeping freight rates high and encouraging congressional opposition to western irrigation development were two of the dilatory tactics employed. The established interests argued that to use federal tax dollars to subsidize the building of western irrigation dams and canals was forcing the eastern growers to subsidize their own competition. About the only counter argument the westerners could muster was that the East had for many years been receiving tax dollars to build harbors and other municipal installations.

In any event, the building of the irrigation systems continued in the West, some by private interests and some by the fledgling U.S. Bureau of Reclamation. And where the water was transported — sometimes in board-lined flumes, sometimes in earthen ditches, and other times in concrete-lined canals — the orchards followed.

The Yakima Valley

In the Yakima Valley, one of the earliest Pacific Northwest regions to begin serious fruit production, the first commercial orchard reported to be set out was in the area known as Parker Bottom — about five miles from where the present city of Yakima is located. H.J. Bicknell was the orchardist. One year later, Fred Thompson set out — in the same locality — what is regarded as the first commercial apple orchard in the valley. The ten-acre orchard was made up of three acres of prunes, three acres of peaches, three acres of apples (Ben Davis variety) and one acre of pears.

Within a couple of years, two irrigation companies were organized. One was for development around Selah, the other around Sunnyside. The latter eventually was taken over by the Reclamation Service.

By 1921, apples were big business. Peaches and pears also were being grown in lesser quantities — about twelve percent of the total fruit acreage of Yakima County.

At that time Winesap was the leading apple, but heavy plantings had been made of Jonathan, Ben Davis, Rome Beauty, Esopus (Spitzenburg), Yellow Newtown, Delicious, Stayman, and Gano varieties. Smaller acreages of Arkansas (Black Twig), Baldwin, Wagener, Grimes Golden, and Arkansas Black existed in the Yakima region.

(Today, about the only varieties of these early choices that a grower might raise would be the Rome Beauty and the Delicious.)

Wenatchee and North Central Washington

As in the Yakima Valley, where pioneer cattleman Ben Snipes once grazed huge herds of cattle, almost the only vegetation in the bottoms and along the benches of the Wenatchee Valley watercourses was sagebrush and bunchgrass. It took the building of diversion dams, ditches, and canals to transform the nearly barren wastes into a lush landscape of fruit trees.

The Gunn Ditch was the first Wenatchee area water diversion, bringing water to six hundred acres in 1896. Next was the Highline Canal, designed to irrgate nine thousand acres of orchards. By 1913 some twenty thousand acres were reported as being moistened with water drawn from the nearby Cascades. The orchard development extended west of Wenatchee — up the Wenatchee River — toward the smaller towns of Molitor, Cashmere, and Leavenworth, as well as farther north to the Chelan and Okanogan area. Before the orchard developers were done, they had extended their plantings far north of the United States-Canadian border.

It should be noted that during this period of orchard expansion, large blocks of land were touted as good apple acreage. But to their sorrow, many investors and would-be growers lost their collective shirts.

By the end of the second decade of the twentieth century, the total acreage in northcentral Washington (including the Wenatchee Valley and the Chelan-Okanogan region) devoted to apple raising was forty thousand acres. This area's growers could rightfully claim that they were producing the nation's highest percentage of extra-fancy and fancy fruit.

Fruit trees flourish downstream from Lake Chelan.

Snipes Mountain orchard is situated in the Yakima Valley.

The main apple varieties grown during this period of time were Winesap, Jonathan, Delicious, Spitzenburg, Stayman, Rome, and Yellow Newtown.

Other Washington Fruit Areas

Elsewhere in the state, more growers hopped on the tree fruit bandwagon, encouraged by the Yakima and Wenatchee successes. The Spokane Valley and the Walla Walla District both had visions of creating their own tree-fruit centers. However, neither area proved to be as well suited.

The Spokane orchards mostly were planted along the Spokane River to the east of the city. As the region grew in population and as small businesses multiplied, suburban sprawl began to replace the orchards.

Oregon

Commercial fruit growing in Oregon, primarily in the Hood River area, began about the same time as it did in the Washington state valleys farther north. However, the climate around Hood River, because it's in the center of the Cascade Range, is considerably wetter than in the Yakima and Wenatchee valleys. This

Commercial orchards pattern the Hood River, Oregon landscape.

Orchards carpet a flat bench in central Washington.

difference in natural precipitation and temperature resulted in lower average yields.

The Hood River Valley, noted for its close-by scenic neighbor, Mt. Hood, does not cover as large an area as does its competitors in Washington. The valley is narrow, varying from two to eight miles in width, and is twenty-four miles long.

Oregon's Willamette Valley, on the other hand, is a large region, covering some ninety miles from Newberg on the north to Eugene in the south and about fifty miles east to west. However, even more precipitation falls on the Willamette Valley. In addition, the heat of the summer and autumn days is trapped between the Cascades on the east and the Coast Range on the west side of the valley. Thus the orchards of this area do not experience the cool nights common to the eastern valleys of the Cascades. Cool temperatures are regarded as necessary for most popular varieties of red apples to acquire the deep color desired by consumers.

Farther south in Oregon, the Rogue River Valley attracted many early would-be orchardists. By the 1920s fruit plantings were common around Medford,

Bluecrop blueberries are grown in in Corvallis, Oregon.

Ashland, Talent, and Phoenix. Commercial fruit plantings included thirteen thousand acres of pears and ten thousand in apple trees. Yellow Newtowns, Esopus (Spitzenburg), Jonathan, and Ben Davis were the most common apple varieties.

Additional, smaller fruit-growing areas were developed around Roseburg and Milton-Freewater, south of Pendleton.

Although Washington orchardists produced more fruit than their colleagues south of the Columbia River, Portland, Oregon took an early lead over Seattle as a fruit shipping center. Sam Birch, Lydia Hall's father (described on page 8), was one of Portland's early fruit exporters to the eastern United States and to Europe.

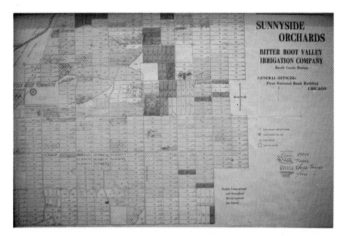

Map of early Bitterroot irrigation project from a development called Sunnyside Orchards shows details of land for sale.

Idaho

The pink and white blossoms of fruit orchards showed up later in Idaho than farther west in Oregon and Washington. Perhaps the early pioneers, with visions of the distant Oregon Country firmly fixed in their heads, kept right on pressing westward through southern Idaho to the Willamette and Columbia rivers. It took a while before they realized that considerable agricultural potential existed in the Boise region they were traversing.

However, by the second decade of the twentieth century, four fruit-growing areas were to be found in Idaho: the Payette and Boise valleys, and the Lewiston and Twin Falls districts. In the Payette Valley, Jonathan apples were the most common, with some plantings of Rome Beauty, Gano, Ben Davis, and Winesap to be found.

The Vanished Lewiston Orchards

Around 1906 it seemed as if the treeless slopes near the northern Idaho town of Lewiston would be a good site for a commercial orchard. A source of irrigation water was not too far away and the sloping terrain offered good nighttime air drainage. Some six thousand acres were developed by a company for sale to investors. As it turned out, however, few trees reached the full bearing stage, although they were planted and cared for on a commercial basis.

The area still is called the Lewiston Orchards — but today one has to look hard to find a block of fruit trees. Suburban sprawl — which is still happening in the Spokane Valley — has eliminated all but a few vestiges of this once-large fruit development.

Montana

Hopes were high at one time for substantial fruit production from Montana's Bitterroot Valley, stretching south some sixty miles from Missoula. More than twenty thousand acres were planted to apples shortly after the turn of the century, but as recounted in the material on the Charles Swansons (page 60), these optimistic plans were dashed by the advent of disease outbreaks, severe competition from Washington and Oregon, and some hard winters.

British Columbia

Fruit growers in British Columbia followed almost the same script as did the orchardists of the United States. In fact, some of the key early players were the same. John McLoughlin, factor of the Hudson's Bay Company at Fort Vancouver on the Columbia River, influenced the agricultural development of British Columbia as well as that of Oregon and Washington. He was later described as a great pioneer in agriculture.

An additional similarity between the United States and the British Columbia developments was the would-be gold miners. Many seekers after the shiny metal came between 1858 and 1860 to British Columbia, as they did farther south. Finding that other fortune hunters were there first, the latecomers turned to the rich soils — at least long enough to raise crops so they could continue their El Dorado hunting! The high yields of vegetables, grains, and fruit convinced some that a more secure future lay in growing crops than panning for elusive gold.

This orchard is located near Othello, Washington.

By 1880, fruit trees were growing in nearly all the settled areas of the Canadian province. Numerous schemes soon emerged, promising nearly instant wealth to be picked from the boughs.

The Duke of Argyll joined in these promotional efforts in 1910 when he wrote:

Southern British Columbia is the finest fruit country on the continent, producing fruit in abundance and of superior quality. In 1891 the total orchard area was 6,431 acres; in 1901 it had only increased to 7,430 acres, but between that and 1904 the total was raised to 13,430 and in 1905 to 29,000 acres.

British Columbia fruit exhibited in England and Scotland carried off the Royal Horticultural Society's gold medal in 1905, and again in 1906, in addition to securing a gold medal at Edinburgh and many prizes at provincial shows.

In the end, however, the British Columbians, like their United States counterparts, found that climate, soil, and water availability were factors that could defeat the most determined orchardist. The lower Fraser River Valley and part of Vancouver Island proved especially suitable for some tree fruits and the berries. But the best place to grow apples, apricots, cherries, pears, and peaches proved to be the semi-arid river valleys across the border from the Washington state fruit-growing areas. The Okanagan, Similkameen, and Kettle River valleys had a good supply of water in the nearby mountains. The climate was not too severe (except for an occasional hard winter) and the soil of the benches and terraces above the rivers and lakes was fertile.

The tree-fruit centers of Penticton, Kelowna, Vernon, Summerland and Osoyoos soon materialized, later to wax and wane in accordance with fruit prices.

As in the United States, many irrigation developments around these towns started as private enterprises. Many had to be taken over by government agencies to insure their continuation.

By the beginning of the 1980s, more than ninety percent of British Columbia's apple crop was being produced in the Okanagan and Similkameen valleys. The chief varieties were Red and Golden Delicious, McIntosh, Spartan, Winesap, and Newtown. Still other varieties to be found included Rome Beauty, Stayman, Wealthy, Jonathan, Yellow Transparent, Duchess, and Tydeman's Red.

It should be emphasized that to fashion a viable fruit industry in the Pacific Northwest took more than fruit trees, water, favorable climate and hardworking fruit growers. The infrastructure of local roads, electrical lines, packing houses, storage warehouses, and transportation networks had to evolve. Fruit on the tree or in the bin had little value by itself.

The cooperative educational and research efforts of the state and provincial agencies and educational institutions also were key components in developing the Pacific Northwest fruit industry we see today.

British Columbia town of Naramata is a fruit-growing center.

*Apples are being grown in Mattawa area orchards, far from
the traditional apple regions of Washington.*

Apples Galore!

"This is going to be the apple capital of the Pacific Northwest!"

That is a pretty strong statement coming from a man living in an area where about all there is to see is dry sagebrush interspersed with thin clumps of sun-bleached grass. The nearby low hills also are totally without trees, the nearest town can only boast of two cafes, and the summer sun is hot enough to fry eggs.

Yet Tom Martin of Mattawa, Washington stands firmly behind his bold assertion. "This area is going to be a force in the apple industry as well as in sweet cherries," he declares.

Since most people would have trouble finding Mattawa on a highway map, Martin's statement might well be doubted. After all, the names associated with fruit growing in the Pacific Northwest are Wenatchee and Yakima in Washington, Hood River in Oregon, and Kelowna, Penticton, and Vernon in British Columbia. They are known throughout the world for their fruit-growing prowess. But Mattawa? "Never heard of it!"

However, a searching look at a Washington state map will show that there is, indeed, a wide spot in the road named Mattawa. Look slightly to the east of the center of the state to where the Columbia River is cutting its way south to the Oregon border. Find where Interstate 90 crosses the Columbia at the tiny hamlet of Vantage. About fifteen miles south of Vantage is Mattawa.

A visit to this potential fruit mecca shows little at present to justify Martin's optimism. The area's only highway — which doubles as the town's main street — obviously will need improving before it can stand up under large numbers of heavy trucks.

Nevertheless, Martin believes he and his wife, Elaine, are living in the Land of Destiny. The one-time veteran of the United States Air Force, who retired to Mattawa in the 1970s, explains why: "Thirteen thou-

Bamboo supplies temporary support to young tree.

sand acres of young apple trees have been planted around here. Five thousand of those acres are beginning to bear. Mattawa apples are going to be more than just a drop in the Pacific Northwest bucket."

It would be easy to ascribe Martin's belief that the apple industry's future lies in the sandy soils around Mattawa to his self-interest. After all, he and Elaine sell real estate and insurance in the area, as well as run the local liquor outlet. The central Washington resident terms himself a "lunch-bucket farmer" because he has to maintain a real estate and insurance office while he raises sheep and cows on twenty-two acres east of Mattawa.

But Martin is not the only one who can see a large complex of fruit orchards, packing and storage plants, and shipping facilities coming into being where now there probably are more jackrabbits per square mile than people. A number of the major orchardists of the Yakima and Wenatchee valleys have made large plantings here in the last several years. According to James K. Ballard, an apple expert from the Yakima Valley, these orchardists saw several advantages to the Mattawa area. "The low elevation there — about 750 feet above sea level — means an earlier spring and warmer temperatures," Ballard says. "The Mattawa apples thus will be ready for market earlier, and command a better price."

New Methods Being Tried

However, according to Ballard, the main attraction of the Wahluke Slope, the area around Mattawa, was the new, unfarmed land that is nearly flat.

The gentle topography meant new, improved methods of raising fruit could be followed on the Wahluke Slope. Dense plantings, elaborate trellises, and mechanized cultivation techniques could be used far more easily than they could in the relatively hilly regions of the Wenatchee and Yakima valleys. Fruit growers also prefer to plant orchards on virgin ground rather than on the site of old orchards. Crippling disease organisms may still be lurking in the old orchard soil.

A related factor spurring the recent plantings around Mattawa was the completion of irrigation canals bringing water from the federal government's Columbia Basin Irrigation Project. The Wahluke Slope was the last of the Project's authorized lands to receive water from the Columbia River, diverted from behind the Grand Coulee Dam, some ninety miles to the north.

Water moderates the air temperature on these central Washington benchlands.

Some earlier orchards in the Wahluke Slope area draw their water directly from the nearby Columbia.

Without this water, Mattawa probably would have continued to be, in Martin's words, "a town that looked like hell." The lunch-bucket farmer adds, "Several years ago we had no cafes and our one service station operator gave 'service' when he felt like it!"

Things obviously are now "a-changing" in Mattawa. Whether the town's future will be as bright as Martin believes may not be known for a decade or two.

A Threat to Established Areas?

The implications of the large new orchards along the Columbia River and elsewhere in the Columbia Basin are serious for the established fruit growing regions. Yakima, Wenatchee, Hood River, and the British Columbia Okanagan area have tremendous investments in orchards, packing houses, and storage facilities. They bill themselves as the "Fruit Bowl of the Nation," the "Apple Capital," and similar appellations. They have a long history of successfully growing and marketing their attractive and tasty products. Yet few knowledgeable orchardists in these areas are snickering at upstart Mattawa.

Grady Auvil certainly is not. One of the Pacific Northwest's premier orchardists and packers (see separate profile of Auvil on page 80), Auvil has put in hundreds of acres of Granny Smith apple trees across the Columbia River from Mattawa. He also grows Grannies and other varieties at his home orchard near Orondo, on the Columbia River above Wenatchee.

"The Granny Smith doesn't need to be grown at as high an elevation as does the Red Delicious," says Auvil. "The Red Delicious needs about 1,200 feet of elevation for it to turn a nice red color. There is no such problem with the Granny because it is a green apple, from start to finish."

As can be seen from the aerial photograph of Auvil's Mattawa area orchard on page 72, he has taken advantage of the gently sloping ground to plant one thousand Granny Smith trees per acre in tight, geometrical hedgerows that offer efficiency in cultivation and harvesting.

Another Wenatchee area orchardist, Roger Wortz, also thinks the traditional apple-growing areas may lose the dominance they have held since the early 1900s. "We will not be able to compete with those big orchards," he says. Wortz and his wife Barbara, who live on Lower Monitor Road above Wenatchee, are

Plentiful apple harvest is ready for shipping.

rightfully proud of their neat but relatively small orchard. "We specialize in Early Red Delicious," he states. "We always get top dollar at the packing house." A lunch-bucket farmer like Martin, Wortz holds down a full-time job as a welder at an aluminum plant near Wenatchee.

Too Many Apples?

These massive new plantings on the Wahluke Slope, as well as the new orchards coming into production on the north side of the Royal Slope, and some located near Quincy and Pasco, Washington, have many industry people worried about overproduction as well as the future of the economies of Wenatchee, Yakima, and the Canadian Okanagan. One prominent Yakima Valley orchardist, after touring the row on row of new apple plantings, told Ballard, "This scares me! We will be filling our apple basket full pretty fast!"

According to Ballard, who has worked for a regional fruit tree nursery firm since his retirement from the agricultural cooperative extension service, a thousand acres of apple trees can produce a million boxes of apples each year, six years after dirt has been tamped around the roots of the fledgling trees. It can be figured on the back of any old envelope that the thirteen thousand acres of new plantings in the Mattawa area thus means thirteen million new boxes of apples will be heading to market before long.

The new plantings, plus all the existing orchards, add up to a lot of green, red, yellow — and mixtures thereof — apples heading to the supermarket bins of the world. Oregon, Idaho, and British Columbia also contribute to this flood of apples, but not to the extent of the Evergreen State. For example, it was estimated in the fall of 1986 that Washington orchardists would pick 70,000,000 apple boxes, each containing approximately 42 pounds of apples. Oregon growers were expected to contribute 2,730,000 boxes and Idaho 2,600,000. The Canadian province of British Columbia, primarily the Okanagan region, was forecast to produce 6,953,000 boxes.

In 1985, according to the International Apple Institute, the United States leaders in apple production were as follows (in bushels): Washington, 48,810,000; New York, 26,667,000; Michigan, 26,190,000; California, 14,762,000; and Pennsylvania, 13,928,000. Oregon and Idaho were far behind with 3,810,000 and 3,119,000 bushels, respectively. The 1986 Washington crop was expected to be larger than the 1985 — up to some 67,000,000 bushels.

Apples move along water-filled conveyer.

All of these colorful and flavorful apples mean a lot of Canadian and United States dollars to their growers, packers, and shippers. For example, the 1984 British Columbia apples were valued at $36,728,000 Canadian dollars. The Washington state crop that year was valued at $331,915,000 United States dollars,

according to the Washington Crop and Livestock Reporting Service.

Some observers believe the price per apple box will tumble as soon as the new plantings hit their stride. Desmond O'Rourke, an agricultural economist at Washington State University (WSU), warns that the coming increase in apple crop size will present formidable problems in both the fresh and processed markets. He points out that while the demand for processed apple juice has been growing rapidly, returns to growers for apples used for juice tend to be low, with competition from imports growing. O'Rourke, who heads a special study group at WSU on agricultural markets, points out that United States demand for fresh Washington apples has been growing slowly. "Export markets, after rapid growth between 1975 and 1980, have stabilized," he also states.

A Red Delicious Surplus

According to the farm economist, the major challenge facing the Washington state apple industry is to develop both domestic and export markets for the very large increase in the fresh pack, especially the Red Delicious, that will be coming from the new plantings. "Even if the domestic market could absorb fifty million boxes, there would be over thirty million boxes available for export," O'Rourke says.

He points out that the domestic sales figure for Washington apples was forty-two million boxes in 1983-1984 and that the export market has never yet absorbed more than thirteen million boxes in any year.

In addition to planting more acres to apples, the region's growers are picking more apples to the acre. "The technology of production has changed in numerous ways that have increased yield per acre," states O'Rourke. There has been a move toward more rapidly bearing and more productive strains planted closer together. In 1969, there were about ninety-nine trees per acre, on the average. The newer plantings are averaging almost three hundred trees per acre.

Orchardists also are just plain doing a better job of growing their fruit, O'Rourke states. "Techniques have been introduced and widely disseminated for reduction of alternate bearing; freeze, frost and wind control; better use of chemicals for thinning, pest control, etc.; improved irrigation methods; more careful management of harvesting; and improved handling."

It might be asked at this point why the great increase in apple tree plantings took place between 1979 and 1984. (The plantings did slow down drastically in 1985.) The attractiveness of the Wahluke Slope and the other new planting areas was a major factor but not the only reason. O'Rourke explains that one or more favorable price years encourage existing growers to expand their orchards and attract new entrants to the apple industry. There is considerable lag — as much as seven years — before the effect of orchard expansions may show up.

Other factors adding to the apple tree boom were, first, that orcharding was becoming popular as a part-time occupation, and second, that tax regulations provided favorable tax benefits to nonfarmers in high tax brackets. "In recent years, a number of partnerships or syndications have been formed to provide such benefits to groups of absentee owners," O'Rourke states. "The expansion of apple acreage in Washington has resulted from the interplay of all these forces: profitability, personal satisfaction, and tax benefits."

The new tax legislation passed by the United States Congress in late 1986 may markedly reduce the fiscal benefits of speculating in new orchards. Whether investors may still find some loopholes whereby they can continue to speculate in orchards remains to be seen. But regardless, thousands of young trees are in the ground, the bees will be pollinating their blossoms, and masses of green, yellow, and red apples will be heading to market. In the fall of 1987, these forecasts of greatly increased yields came true. In Washington, the crop was seventy million boxes and, at least initially, the price being paid growers was way down.

Some orchardists, however, are not ready to concede the struggle to the big operators. Tom Perkins, with his twenty-six acres of orchard near Sedro Woolley, was planning to put still more acres into orchard. He says, "The little grower will just have to think smarter!" Perkins advocates growing varieties different than the ones the big operators will be trying to market. "Small growers doing so and selling to the local markets will do better than the large orchardist selling to the national market," he states. Fred Westberg, retired manager of the Washington State Fruit Commission, adds, "Some of the Johnny-come-lately growers are going to find that successful fruit production involves more skill, knowledge, and patience than they are prepared to give."

Ballard also thinks that the smaller operators can keep their niche in the total production scheme. "Some of the highest-quality fruit comes from family-size orchards, about thirty acres in size," he says.

Sampling's half the fun of picking the apples in this flourishing Montana orchard.

The extremely small orchards in Japan — some as tiny as one acre — would seem to support Perkins and Ballard in their conclusions.

Innovation Helps

Some smaller-scale growers are banding together to promote their U-Pick and orchard-direct fruit sales. They jointly announce their orchard locations in newspaper announcements and in printed brochures. They also participate in local farmers' markets.

These innovative operators know they have a potent weapon in the ripeness of their fruit, which they allow to hang longer on the tree before picking. Fruit destined for shipment to a faraway supermarket must be harvested while nearly green.

Older citizens are proving good customers for these down-on-the-orchard sales. They like to get out in the country for a drive to the orchard, they enjoy a chat with the orchardist, and they remember how good fruit used to taste in the pre-supermarket days!

The more enterprising of these regional associations put on special events and fairs to call attention to their ripening apples, peaches, and other fruit. Music and antique sales are featured along with a plentiful supply of fruit pies and other baked goods to remind the visitors of fruit's delights.

Up With Apples, Down With Munchies

The apple growers of the Pacific Northwest would like people throughout the world to eat more apples and less snack food. "Our primary competition is not other apples," a Washington State Apple Commission spokesman was quoted as saying in the October 1986 issue of *Fruit Grower.* "It's the Twinkies and Ding Dongs and Doritos and whatever."

The Washington state promotional organization calculates that the present per capita apple consumption of seventeen to eighteen pounds yearly will have to increase to twenty-three pounds to use up the growing number of apples coming from orchards in Washington and elsewhere in the United States. "We need to inject some excitement back into apples and give people new reasons to eat apples."

Estimates from the U.S. Department of Agriculture are that the per capita consumption of fresh fruit yearly in the United States is right at ninety pounds. Canned and dried fruit consumption amounts to twelve pounds yearly while about thirteen and a half pounds of frozen fruit and juices are eaten or drunk each year by the average person.

One of the ways in which the apple industry is trying to induce more people to eat more apples is to piggyback their product onto the increasing public interest in better nutrition. Apple growers in particular are fond of claiming that eating their product is very good for one's health.

Controlled Atmosphere (CA) Storage

In the 1950s, fruit packers began to dramatically extend the life of their apples by depriving them of oxygen once they had them stored away in their warehouses. Horticulturists had found out through experimentation that the apples could be put into "hibernation." When placed in tightly sealed chambers, cooled to 31°F., and with ninety-five percent of the oxygen removed, the apples essentially stopped maturing. When the fruit was removed some months later from the chambers, it was like fresh-picked apples in flavor and crunchiness

The fruit industry quickly realized that controlled atmosphere (CA) storage was a good idea. By 1983, Washington apple growers alone had room for more than fifty-four million boxes in controlled atmosphere rooms. The use of CA has extended the period of fresh fruit sales of apples to almost fifty-two weeks a year.

Where Does the Fruit Go?

Pacific Northwest fruit, especially the apples, is available for sale in many areas of the world. Accord-

Sorter carefully checks apples in the packing house.

ing to the Washington State Apple Commission, Los Angeles residents buy the most, followed by New York City and Chicago citizens. Around one-quarter of the state's Delicious apples and other varieties go abroad to such customers as Taiwan, Saudi Arabia, Hong Kong, and Canada. Other major overseas customers include Colombia, Singapore, Norway, Sweden, New Zealand, Malaysia, Thailand, the United Arab Emirates, Finland, and the United Kingdom.

When it is considered that seven million tons of apples are grown annually in Europe, it is apparent that Pacific Northwest apples are able to hold their own in world markets. Len Wooton, an apple industry leader from Wenatchee, ascribes the success of the region's fruit growers to the ability of the industry to develop new techniques and to get them from the laboratory to the grower; the strict grading of the fruit; and the lack of restrictive government controls.

Pacific Northwest growers are confident they could even do better in world markets if some countries removed their import restrictions. A number of countries that formerly imported fruit now limit shipments from the Pacific Northwest in order to save their foreign currency reserves. Egypt, Indonesia, the Philippines, and Venezuela are in this category. Other countries, such as Germany and Japan, claim to be restricting imports in order to keep fruit pests outside their own borders. Ballard says, "Japan restricts the importing of Washington apples by claiming they do not want codling moth brought in with the imported apples. However, they already have codling moth infestations."

In another example, pears from the United States cannot be transported into the Scandinavian countries

Fruit areas turn the banks of the Snake River green.

of Finland, Norway, and Sweden until all the locally grown varieties have been sold.

Not a Two-Way Street?

Such restrictions on selling abroad are difficult for the Pacific Northwest to take gracefully — especially since foreign growers are selling their fruit in Pacific Northwest stores. Imports of fresh apples into the United States have almost doubled since 1980, with most of the increase coming within the past three years, O'Rourke reports. Similar import jumps have been taking place with pears and with fruit juices.

British Columbia orchardists, for their part, are hopeful that they have found a new market for their fruit — across the Pacific Ocean in mainland China. The Chinese government recently announced its intention to import sixteen percent of the nation's food. A British Columbia marketing organization would like to capitalize on this intention. A spokesman calculated that if just one Chinese ate one apple per year, they would consume more than twice the present British Columbia production!

Canadian growers are aware of the need for improvement in their apple marketing and production methods. "We have too many old orchards with low-density plantings," says E.R. Hogue, a research scientist at the Summerland Research Station in British Columbia. "We should have ninety percent high-density plantings," he states. "But we are nowhere near that."

The specialist in pomology and viticulture reveals that many British Columbia orchardists, like their United States counterparts, are carrying a heavy debt load. Gerald Geen, president of the British Columbia Fruit Growers' Association, was quoted in a September 1986 Summerland, report as saying, "If returns do not improve dramatically in 1986, a substantial decline in the acreage of tree fruits is predicted for the remainder of the decade. The tree fruit industry of British Columbia is engaged in a desperate struggle."

PROFILE:

A Return to Flavor

Grady Auvil

Maybe you have begun to doubt that fruit ever really tasted as good as you remember it tasted years ago.

If your palate is not quite resigned, however, to mediocre produce without flavor, color, or aroma, then take heart! There's a man in central Washington state whose goal is to return something to the modern consumer that has been taken away over the years — quality produce, especially quality peaches, apricots, cherries, and apples.

"When the 'Mom and Pop' grocery stores were edged out of business by the big supermarket chains," says orchardist Grady Auvil, "an important contact was lost between the producer and the consumer. I've fought this battle for a long time now, the battle to get quality back on the shelves, and I'm beginning to make some headway."

Auvil grew up in the small town of Entiat, Washington, twenty miles north of Wenatchee. Fruit growing was the only way to make a living there at that time. The son of fruit growers Lew and Ida Auvil, he knew from an early age that he wanted to own an orchard. In 1928, just before the Depression, he and his two brothers bought a small piece of land in Orondo, above Wenatchee.

They immediately planted twenty acres of orchard. Then they got busy holding down as many jobs on the side as they could manage in order to keep their land and fruit trees. Today the land holdings total six hundred acres, mostly in Orondo and beside the Columbia River south of Vantage.

Auvil sells his graded fruit entirely to the fresh fruit industry; the Red and Golden Delicious apples he sells through a cooperative and the rest he sells himself, mostly to small chain groups and to street brokers in United States cities. "America is crazy for good fruit," Auvil reiterates. "I sell a lot of my peaches now in California, and almost all of my nectarines. I've never seen a market so keen to get something good to eat being so poorly served by supermarkets. People know quality when they get it and they're beginning to make themselves heard."

Orchards in the narrow Entiat Valley are known for fine fruit.

One of Auvil's favorites is the Granny Smith apple. In fact, Auvil was largely responsible for the shiny green New Zealand native's popularity in Washington state apple country.

As he explains, "I have never encountered an apple with so many pluses and so few minuses. The Granny Smith doesn't need a high elevation and it doesn't need to get red. It can withstand heat in excess of 100°F., and cold temperatures dropping to 15° below zero."

Although Auvil has been in the orchard business for over sixty years, he does not rely on experience and common sense alone. He is constantly experimenting with newer varieties of apples, particularly a number of promising cultivars from Japan, New Zealand, and New York. Two of these are the Gala, a relative newcomer, and an even newer cross between the Jonathan and Golden Delicious, the Jonagold.

A generous supporter of work being conducted at the Tree Fruit Experiment Station in Wenatchee, and by agricultural researchers at Washington State University in Pullman, Washington (two hundred miles east of his orchards), Auvil keeps close tabs on the newest discoveries about optimum storage conditions for his valuable crops. Last year the veteran orchardist made approximately one thousand dollars per acre on his six-hundred-acre business. Auvil finds that when he speaks about fruit growing, people listen.

Apricots, Berries . . . and All the Rest

It could be easy to overlook the rest of the Pacific Northwest's fruit industry — the apricots, the berries, the cherries, the grapes, the peaches, the pears, and the plums/prunes. The apple growers and their marketing organizations make the most noise in the media, the apple orchards are highly visible in many river valleys, and the neat geometric piles of their bright red, green and yellow fruit dominate the produce sections of supermarkets. Yet the region's other fruits, especially the cherries and pears, are major industries in their own right. For example, apricot growers in Washington produced 3.3 million dollars worth of the bright orange fruit in 1986. Sweet cherries brought fifty-nine million dollars to the same state's orchardists that year, while peach and pear production was valued at almost nine million and seventy-two million dollars, respectively. Plum and prune production in 1986 was considerably smaller in value.

Freshly picked cherries are carried in canvas bags.

Apricot Production Stable

There has been some slight growth in apricot production in Washington during the past five years, according to Ken Severn, manager of that state's Fruit Commission. "The apricot industry has been stable for the last fifteen to twenty years," he says. "Production recently has been ranging from three to four thousand tons per year."

These totals are tiny compared to what California growers pick off their graceful trees. In 1985, the California orchardists were expected to produce 128,000 tons of apricots. The Washington Crop and Livestock Reporting Service states that orchardists in the Golden State produce ninety-five to ninety-seven percent of the nation's apricot crop.

Severn reveals that Washington apricot orchardists sell one-third of their crop in the local region with the balance being shipped to national markets.

Sweet Cherries Doing Well

Washington sweet cherry growers, like their apple brethren, are doing well in the United States market. In 1985-86, Washington orchardists produced forty-five percent of the nation's sweet cherries. According to Severn, Washington's Yakima Valley is the locale of much of this sweet cherry production, primarily the

Bing, Black Republican, Chinook, Lambert, Rainier, Royal Ann, and Van varieties.

Some sweet cherries are also grown in Montana in close proximity to that state's Flathead Lake, and in Idaho's Boise Valley. According to Severn, few tart (sour) cherries are grown in Washington.

British Columbia orchardists produce sizable crops of cherries, primarily in the Okanagan Valley area close to the region's lakes. At the beginning of the current decade, some 2,800 acres of sweet cherry trees and 200 acres of sour cherry trees were being grown.

Oregon cherry orchardists are substantial producers of the tasty fruit. In 1985, some 29,000 tons of sweet cherries were grown in Oregon, along with six million pounds of the tart varieties.

Peaches—Good Flavors But Winter-Tender Trees

Peach orchardists in the Pacific Northwest can grow the same varieties as their California counterparts. Some observers rate the flavor of the Pacific Northwest specimens as superior to the California fruit. However, the more severe winters, especially in Washington, keep the regional peach production low. According to state official Severn, Washington peach growers are likely to produce a regular-size crop only three years out of five. In the other two years, the crop will be lost to winter damage to the blossoms or to the trees themselves. "The cold may cause the tree to be killed or limb loss to occur," Severn says. "Growers are unable to keep a good scaffolding structure in their trees because of the winter limb losses."

In spite of these climate difficulties, the annual production of peaches in the Pacific Northwest is substantial. Idaho growers picked eleven million pounds of peaches in 1985, Oregon orchardists contributed fifteen million pounds, while Washington producers boxed some twenty-nine million pounds of the bright-yellow-to-red fruit. In Washington alone, the 1986 peach crop was valued at 8.7 million dollars.

Pears — Winter and Summer

A recent newspaper article about pears by a food editor stressed what the editor thought was the sensuous nature of this fruit. Pear-shaped, lush human forms, singer's pear-shaped tones, and prized pear-shaped diamonds were described, compelling the writer to conclude that a pear might be the world's "most seductive fruit, even more sensuous than Eve's tempting apple."

Whether such considerations enter into shoppers' minds when they pick up a supply of Bartletts, Boscs, Seckels, or one of the other pear varieties is unknown. However, there is no question that consumers buy a lot of Pacific Northwest pears. In 1986, Washington state pear growers, for example, were estimated to have deposited 266,000 tons of the "sensuous fruit" in their orchard bins. Greenish-yellow Bartletts made up about half of this total. Oregon growers were expected to produce 193,000 tons that year, with the Bartletts making up better than one-third of the total.

California, as might be expected, is the major state in pear production, with some 292,000 tons — virtually all Bartletts — produced in 1984. The entire United States production of pears does not match up to one European country's, however. The United States total production of pears in 1985 was expected to be 678,000 metric tons, whereas the production of Italy's pear orchardists was calculated to be almost one million metric tons.

The canning of pears is a sizable industry in Washington, Severn points out. "Processing uses eighty percent of the Bartlett crop in Washington," Severn states. "There are three pear canneries in the Yakima Valley, two in Oregon, and two in California."

According to Ken Severn, one of the California canneries is producing only limited amounts of canned pears.

Plums and Prunes

The U.S. Department of Agriculture reports the plum and prune production for only four states out of the entire fifty. However, three of these four states are Idaho, Oregon, and Washington, thus indicating the importance of the Pacific Northwest to this segment of the fruit industry. Michigan is the other state represented in the statistics. Washington and Oregon compete for the lead in plum and prune crops, with the winning totals shifting back and forth. In 1985, for example, Oregon orchardists picked 25,000 tons of the blue-red fruit. Washington growers had to be satisfied with 10,200 tons. The previous year, Washington orchards yielded 18,000 tons and Oregon's 15,000 tons. It should be noted, however, that California-grown plums can be found in Pacific Northwest supermarkets nearly all summer long.

Crew packs berries at Weiss Ranch, Snohomish Valley.

Ken Severn reports that the production of dried prunes has been steadily declining for a number of years. "Some two to three thousand tons are produced yearly in Washington," he states. He ascribes the decline in interest in dried prunes to the modern trend of eating fresh fruit.

It's the Berries

Blackberries, blueberries, cranberries, raspberries, and strawberries — these delightful products are generally referred to as the small fruits. However, despite their size relative to the apples and pears, there is nothing small about the economic importance of berries or the eating pleasure they give. In Washington, for example, the value of the 1986 crops of these five different berries amounted to over twenty-five million dollars. Red raspberries led the way in value in this state, with strawberries and cranberries coming in second and third, respectively. It is about the same story in British Columbia to the north. In 1982, the province's berry farmers received over 23 million dollars (Canadian) for their raspberries, 10 million dollars for their strawberries, 7 million dollars for their cranberries, and 6.3 million dollars for their blueberries. At that time a major expansion in cranberry bogs was being planned by the British Columbia growers.

Cranberry growers in the Pacific Northwest cannot claim to lead the United States and Canada in numbers of these holiday delights, but they can brag that their berries are redder! The major producers of cranberries in the United States are located in Massachusetts (11,300 acres in 1985), Wisconsin (7,500 acres), and New Jersey (3,300 acres). In that year, Washington cranberry bogs covered 1,200 acres while Oregon's total was 1,000 acres. However, the climatic conditions along the Washington and Oregon coasts, where the states' bogs are located, result in redder berries. Consequently, the major distributors of cranberries in the United States blend some of the western berries with the lighter-colored eastern berries to induce supermarket shoppers to add cranberries to their shopping carts.

The big story in Pacific Northwest berries in recent years has been how strawberry growers in California and Mexico have been taking over the market. In 1966, California, Oregon, Washington, and Mexico strawberry growers supplied ninety percent of the United States market, with each state and Mexico supplying about an equal share. By 1982, however, growers in California had captured seventy percent of the market. Oregon growers saw their share decline to seventeen percent. By 1985, California strawberry fields were producing 334 million dollars worth of the plump berries whereas Oregon growers had to be satisfied with 15 million dollars. Washington farmers were a distant third with 6.9 million dollars.

Oregon growers for some years blamed their state's child labor prohibitions — reportedly not matched in

Harvester and catch bin gather ripe grapes.

California or Mexico — for their decline in national strawberry sales. However, extensive research into the problem concluded that aggressive innovation on the part of the California strawberry industry was a more likely cause. Better varieties, making possible an extended growing season (ten months compared to one to two months in Oregon), higher yields per acre, and production for both fresh and frozen sales were the major California strategies that netted such large market gains.

Grapes, Wine and Otherwise

It has been repeatedly reported that the grape industry in the Pacific Northwest has been on a growth roll during the last decade. There have been some declines and flat spots but the general thrust has been to more plantings. This growth has been in the face of the once-held belief that wine grapes were the "private property" of California growers in that state's Napa and San Joaquin valleys.

It is still true that California viticulturists dominate the national market, especially in raisin and table grapes. A glance at the United States production figures should be convincing. In 1985, for example, California vineyards produced over 5 million tons of grapes. Raisin varieties made up nearly half of this total. New York State was in distant second place, with 146,000 tons. Washington vineyards contributed 116,100 tons to the national pile of grapes. Data for Idaho and Oregon are not reported by the U.S. Department of Agriculture.

In Washington, the big grape is the Concord. This pleasantly aromatic juice grape accounts for about ninety percent of the state's total grape production. However, there has been little growth in Concord acreage for several years.

The opposite is true for wine grapes. In 1987, it was reported that 11,000 acres in Washington were planted to different varieties of wine grapes. In 1983, the total was only 8,000.

This growth in wine grape acreage has been matched in numbers of wineries. In 1986, Washington state had fifty-four wineries, a number closely matched by Oregon with its fifty operating wineries. In 1984, a book on Canadian wineries listed twelve wineries with a British Columbia address, mostly in the Okanagan region.

The Pacific Northwest wineries and vineyard operators have reason to be proud of the quality of

Forklift operator tips grapes into large bin.

Auger moves grapes along to crushing operation.

their offerings, with numerous gold, silver, and bronze medals being awarded to them at international competitions. However, they probably wish the per capita consumption of wine in the United States and Canada would increase to the levels of some countries. The disparity is surprising. The citizens of France consumed nearly 24 gallons each in 1981, followed by the Portuguese and Argentinians with nearly 20 gallons each, and the Italians at 19½ gallons each.

In contrast, United States residents put away a little over 2 gallons yearly. Citizens of the United Kingdom drink 1.7 gallons yearly.

PROFILE:

The Success of Aplets and Cotlets

Fruit confections are made from apples and apricots.

Ask a Washingtonian about Aplets and Cotlets: If you get a blank stare, you're speaking to an imposter! Aplets and Cotlets candies are to the apple country of central Washington what sourdough bread is to San Francisco.

The Liberty Orchards Company of Cashmere, Washington, founded by Armenian immigrants in 1920, has been producing the tempting fruit-flavored candies for over sixty-five years. Gregory Taylor, grandson of one of the company's founders, is president of the company today and personally supervises the candy making in the small but highly successful company.

When Taylor's ancestors, Armen Tertsagian and Marcar Balaban, first came to the United States, they started in Seattle, where they opened an Armenian restaurant. That failed so they tried a yogurt factory, but that didn't fare any better. When they traveled to the Wenatchee-Cashmere area they immediately fell in love with the country, which closely resembled the land they were born in. Still trying to start a business, they capitalized on the fact that the market for apples was not good — there was an excess of apples, and most of the farmers and orchardists were underfinanced.

Attempting to create a use for the apple, the resourceful Armenians experimented with apples in a Near Eastern confection called "Rahat locum," or "Turkish Delight." The original Turkish Delight, according to Taylor, consisted of a jellied candy flavored with exotic and aromatic ingredients like rose water and essence of orange.

Substituting apple juices for the original flavors, the men arrived at a succulent, jellied candy much like its foreign ancestor — Aplets was born. This candy was followed shortly by Cotlets, then by Grapelets, and, now, by a new assortment of fruit and nut combinations called "Fruit Festives."

At first the confections were sold locally, then only in West Coast stores. Today more than 150,000 customers regularly buy Aplets and Cotlets.

According to Taylor, the two founders, grateful to their adopted country for offering them freedom from racial persecution, named their original business, an orchard and cannery, "Liberty Orchards." The story is a classic immigrant story, Taylor says. "The land of opportunity gave them a chance to excel and they were determined to do it!"

Fruit Stands

Colorful streamers and a red-and-white umbrella brighten this roadside fruit stand.

No discussion of the fruit industry in an area would be complete without at least mentioning a sight familiar to anyone who's ever taken a trip on a major roadway in the spring or summer — the roadside stand.

Such stands range from a simple table and umbrella with a few bushels of produce picked fresh that day, to a "super stand" like the one owned by Bob and Claudia Spanjer of Cashmere, Washington.

The Spanjers opened Bob's Apple Barrel on U.S Highway 2 in 1977. "I owned orchard land right beside the highway at the time anyway, and I kept thinking that a fresh fruit stand would just be a natural," Bob said. The family relies on plenty of good fruit at reasonable prices, easy parking, room to browse, and a long season in order to attract a large clientele.

"I'm not afraid of competition," Spanjer says, grinning. "I know if I can provide better produce and better service than other stands, I'll get the customers." He added that most of his customers come from the larger coastal cities like Seattle. These people are out for a Sunday drive or a summer trip and know they can stop at Bob's for fruit immediately before they head for Stevens Pass and the west side of the mountains. "I keep up on the going prices for fruit in the valley and mark mine accordingly. I more or less set the prices for fresh fruit sales in the valley," Spanjer maintains.

One of Bob's specialties is a fresh fruit milkshake, made in a kitchen to the rear of the fruit stand while you wait. You specify the fruit. But, a word to the wise — you won't be ready for a meal for several hours after consuming one of these delights. And, when you order one from Bob or one of his employees, ask for a spoon!

Remember "Tiny's"?

The Spanjer's roadside fruit stand is right across the highway from the spot where the famous "Tiny's" was located from 1953 to 1972. Like its distant cousin, Wall Drug in Wall, South Dakota, Tiny's was heralded by signs that began hundreds of miles on either side

A careful shopper compares two nectarines.

Boxes and boxes of ripe apricots are proudly displayed.

of it. The signs themselves became a part of western Americana.

In fact, Tiny himself was a bit of a landmark. Until his death in 1972, Tiny (Richard Duane Graves) was, his sister, Sharon Hill, recalls, "Generous to a fault, big-hearted, kind and giving, and creative." The fruit stand was the result of his innovative thinking, Hall added. "Tiny had given a lot of thought to what would attract people, and what would make them keep coming back. He had a knack for appealing to peoples' interests in the unusual, the out-of-the-way, the quaint."

Appetizing fruit comes in serve-yourself baskets.

Three kinds of apples are featured at this roadside fruit stand.

"Tiny's was such a favorite stop," Hall said, "that friends would ask if they could take a sign and put it up somewhere around their own hometown."

The Family Fruit Stand

Other fruit stands in the area are run by entire families. Some have been in those families for several generations. The fruit stand operated by Della and Jack Feil (Feil Orchard and Fruit Stand) near Orondo, Washington, has been in business in the same location for sixty years. They have been allowed to keep their choice location right next to the highway, Jack says, because they existed in that location before a city ordinance required that fruit stands be set back a number of feet off the highway.

Feil has no children to take over the stand, so it will pass out of existence when he and his wife retire.

So, whatever your tastes might run to — fresh fruit from a roadside stand, one of Bob Spanjer's miraculous mixtures of fruit and ice cream, or any of the myriad new fruit product snack foods — go ahead, indulge. It's good for you. And, oh, yes, pass the fruit, please!

PROFILE:
Fruit and Health

Helen Mitchell of Spokane enjoys an apple.

It's a hot summer evening and you're seated on your patio, recuperating from that second helping of mashed potatoes and an extra barbecued pork chop that you really didn't need. A cool and flavorful dessert of some kind would be nice, you think, but ice cream is not exactly what your palate, or your stomach, craves. Enter a variety of new fresh-fruit and fruit-juice-flavored frozen desserts that would tempt even the most satiated gourmand.

The choice of fruit in these new concoctions is yours: peaches, raspberries, strawberries, pineapple, bananas, blueberries, and all of the above — with or without cream. Smoother than the sherbets of the recent past and less than half the calories of most regular ice creams, these new desserts have been an almost instant success.

Curt Sturm, General Mills sales representative for an area from eastern Montana west to the Cascades and south to Oregon, says that the new dry fruit snack foods are increasingly popular, too.

Gorgeous fruit pleases the eye (and the palate) at Seattle's Pike Place Market.

Fairgoers look over the fruit display at a Portland, Oregon fruit show.

Carefully selected apricots are weighed at the farmers market in Pasco, Washington

"There's no doubt at all that the fruit bars and fruit bits are category leaders in the 'snacks and portable foods' market area for General Mills," Sturm says. According to Sturm, granola bars and other similar products led the pack for snack foods until recently. "Now most strength is in the fruit products; not only are they competing well with items like granola bars and other high-energy-type bars, but they are competing well in a highly competitive category," Sturm adds.

Food researchers say, in fact, that since about 1980 there has been a definite shift in peoples' tastes away from canned and processed fruits and vegetables to fresh. While the trend has spelled doom for nearly thirty canneries in California alone in the last five years, it has rejuvenated small produce farms, farmers' markets, and roadside fruit stands.

The change is attributed by some to a growing health consciousness on the part of the American populace. Also, thanks to improvements in trans-portation and to growing methods, people can get high-quality fresh produce from the farmer down the road or from halfway around the world — with nearly equal ease.

It is common to see supermarket bins filled with crisp and colorful fruit that was picked just a few days earlier on another continent. We think of the jumbo jets speeding northward from Central and South America as bringing the exotic fruits of the tropics, yet these huge airplanes also deliver more conventional crops of plums, peaches and grapes to the United States consumer.

According to Ben Wong, an economist for the U.S. Department of Agriculture, Americans' per capita consumption of canned fruit declined by almost half between 1980 and the end of 1985, from 16.2 pounds to 9.5 pounds. Consumers have not given up fruit, however. Now they want fresh. And orchardists and fruit stand operators in the Pacific Northwest are ready to give them what they want.

Roses bloom near 140-year-old Black Tartarian cherry tree in Eugene, Oregon.

Bibliography

Books

All About Growing Fruits & Berries. San Francisco: Ortho Books, Chevron Chemical Company, 1982.

Apple Growing in the Pacific Northwest. Portland, Oregon: Young Men's Christian Association, 1911.

Baker, Harry. *Simon and Schuster's Step-by-Step Encyclopedia of Practical Gardening: Fruits.* London: Mitchell Beazley Publishers, Ltd., 1980.

Carlson, R.F., et. al. *North American Apples: Varieties, Rootstocks, Outlook.* Lansing: Michigan State University Press, 1970.

Childers, Norman F. *Modern Fruit Science.* Sixth Ed., Horticultural Publications, New Jersey: Rutgers University, 1975.

Duke of Argyll. *Yesterday and Today in Canada.* London: George Allen & Sons, 1910.

Folger, J.C., and Thomson, S.M. *The Commercial Apple Industry of North America.* The Rural Science Series, L.H. Bailey, Ed. New York: The MacMillan Company, 1921.

Fraser, Samuel. *American Fruits.* New York: Orange Judd Publishing Company, Inc., 1927.

Hedrick, U.P. *The Peaches of New York.* Twenty-Fourth Annual Report, Vol. 2, Part II. New York: J.B. Lyon Company, 1917.

Hedrick, U.P. *Grapes and Wines from Home Vineyards.* England: Oxford University Press, 1945.

Hill, Lewis. *Fruits and Berries for the Home Garden.* New York: Alfred A. Knopf, 1977.

Kraft, Ken and Pat. *Fruits for the Home Garden.* New York: William Morrow & Company, Inc., 1968.

Paddock, Wendell and Whipple, Orville B. *Fruit Growing in Arid Regions.* The Rural Science Series, L.H. Bailey, Ed. New York: The MacMillan Company, 1912.

Robinson, Jancis. *Vines, Grapes, and Wines.* New York: Alfred A. Knopf, 1986.

The Royal Horticultural Society and Geoffrey Cumberlege. *The Fruit Garden Displayed.* London: Oxford University Press, 1951.

Shoemaker, James S. *Small Fruit Culture.* Fifth ed. Westport, Connecticut: The AVI Publishing Co., Inc., 1978.

Shortt, Adam and Doughty, Arthur G. *Canada and Its Provinces.* Vol. XXII. Glasgow: Brook and Company, 1914.

Taylor, Griffith. *Canada.* London: Methuen & Co., 1947.

Tukey, H.B. *The Pear and Its Culture.* New York: Orange Judd Publishing Company, Inc., 1928.

Walheim, Lance and Stebbins, Robert L. *Western Fruit, Berries, and Nuts: How to Select, Grow, and Enjoy.* Tuscon, Arizona: H.P. Books, Inc., 1981.

Warkentin, John, Ed. *Canada, A Geographical Interpretation.* Ontario, Canada: Methuen Publications, 1968.

Weaver, Robert J. *Grape Growing.* New York: John Wiley & Sons, 1976.

Woodcock, George. *The Canadians.* Cambridge, Massachusetts: Harvard University Press, 1979.

Pamphlets

Adams, E. Blair. *Small Fruits for Home Gardens.* Extension Bulletin 0708, Cooperative Extension, College of Agriculture and Home Economics. Pullman, Washington: Washington State University, 1984.

Ballard, James K. *Granny Smith, An Important Apple for the Pacific Northwest.* Extension Bulletin 0814, Cooperative Extension, College of Agriculture and Home Economics. Pullman, Washington: Washington State University, 1981.

Basic Care for the Home Orchard. Clackamas, Oregon: Home Orchard Society, 1986.

Bitterroot McIntosh Recipes. Hamilton, Montana: The Bitterroot Valley Historical Society, 1981.

Commercial Red Raspberry Production. PNW 176, Washington, Oregon, Idaho. Edited by

William P.A. Scheer. Pullman, Washington: Washington State University, 1987.

Fruit Tree Cultivars in British Columbia. Publication 1609. Agriculture Canada, 1977.

Lamonte, Edward R. and O'Rourke, A. Desmond. *Red Raspberry Industry in the Pacific Northwest.* EB1333, Agricultural Research Center. Pullman, Washington: Washington State University, 1985.

Norton, Robert A. *Tree Fruit Cultivars for Western Washington Homes and Orchards.* Extension Bulletin 0937, Cooperative Extension, College of Agriculture and Home Economics. Pullman, Washington: Washington State University, 1982.

O'Rourke, A. Desmond. *The Future Size of the Washington Apple Crop.* Provisional Impact Center Working Paper No. 1. Pullman, Washington: Washington State University, 1984.

Swales, J.E. *Commercial Apple Growing in British Columbia.* Province of British Columbia: Ministry of Agriculture and Food, Reprinted 1982.

Pruning the Home Orchard. Extension Bulletin 0660, Cooperative Extension, College of Agriculture and Home Economics. Pullman, Washington: Washington State University, 1985.

Washington Agricultural Statistics, 1984-85 and 1985-86. Olympia, Washington: Washington Crop and Livestock Reporting Service, October 1985.

Catalogs

Bear Creek Nursery Fall '86 - Spring '87, Northport, WA.

Fruit Tree Guide, May Nursery Company, Yakima, WA.

Hilltop 1987 Catalog and Handbook, Hartford, MI.

Van Well Nursery, Wenatchee, WA.

Yakima Valley Nursery, Yakima, WA.

Index

Authors David C. Flaherty and Sue Ellen Harvey stand next to the plane used to take many of the outstanding aerial photographs in this book. David was educated at Stanford University and has written numerous books, articles, and papers as well as writing, photographing, and directing films. He recently retired from Washington State University's Engineering Extension Service. Sue Ellen is currently affiliated with Washington State University (where she obtained her master's degree) and has been a technical editor and an English instructor. Her other publications include an English textbook, a journal of scholarly activities at WSU, and numerous newspaper articles as well as poetry.

More Garden Reading

Pacific Northwest Gardener's Almanac gives detailed information on growing vegetables and herbs between the east slope of the Rocky Mountains and the Pacific Ocean, the Rogue River Valley and central British Columbia. Author and Master Gardener Mary Kenady tackles Northwest gardening with proven techniques and experience.
($14.95 / $18.95 Canadian, softbound, illustrated, 192 pages).

Alaska's Farms and Gardens provides an in-depth look at the crops and agriculture of the North. It includes a review of early agriculture in Alaska and a map showing both present and potential farm and pasture lands.
($12.95 / $16.95 Canadian, softbound, color, 142 pages).

If you like learning about wild plants, these books will interest you: *Alaska Wild Berry Guide and Cookook* ($14.95 / $18.95 Canadian, softbound, color, 200 pages), *Alaska-Yukon Wild Flowers Guide* ($16.95 / $21.35 Canadian, softbound, color, 218 pages), *Alaska's Wilderness Medicines* ($9.95 / $12.65 Canadian, softbound, illustrated, 100 pages), *Plant Lore of an Alaskan Island* ($9.95 / $12.65 Canadian, softbound, illustrated, 194 pages).

Ask for these books at your favorite bookstore, or order directly from the publisher. (Add $1.50 per book to cover postage and handling):

Alaska Northwest Publishing Company

130 Second Avenue South, Edmonds, Washington 98020

or call toll-free 1-800-533-7381 to use your VISA or Mastercard

A full catalog of other fascinating books is available